Clinical Essentials of Calcium and Skeletal Disorders

First Edition

Leonard J. Deftos, MD, JD
Professor of Medicine
University of California, San Diego
Staff Physician and Head of the
Endocrine Research Laboratory
San Diego VAMC
La Jolla, California

AF882182

Professional Communications, Inc. A Publishing Corporation

Copyright 1998
Leonard J. Deftos, MD, JD

Published by:
Professional Communications, Inc.

All rights reserved. No part of this publication may be reproduced or transmitted in any form or by any means, electronic or mechanical, including photocopy, recording or any other information storage and retrieval system, without the prior agreement and written permission of the publisher. Requests for permission or further information should be faxed to Professional Communications, Inc. at 580/367-9989, or mailed to P.O. Box 10, Caddo, OK 74729.

For orders, please call:
1-800-337-9838

ISBN: 1-884735-39-8

Printed in the United States of America

DISCLAIMER

The opinions expressed in this publication reflect those of the author. However, the author makes no warranty regarding the contents of the publication. The protocols described herein are general and may not apply to a specific patient. Any product mentioned in this publication should be taken in accordance with the prescribing information provided by the manufacturer.

This text is printed on recycled paper.

DEDICATION

This book is dedicated
to the memory of my son
Stephen Minot Deftos.

ACKNOWLEDGMENT

The author has been generously supported by the Department of Veterans Affairs and the National Institutes of Health.

ΑΦΙΕΡΩΜΕΝΟ ΣΤΟΝ ΕΑΥΤΟ ΜΟΥ

TABLE OF CONTENTS

TABLES

FIGURES

About the Author

Leonard J. Deftos, MD, JD, is a Professor of Medicine at the University of California, San Diego, Head of the Endocrine Research Laboratory of the San Diego VA Medical Center, and the President of the San Diego Bone Club. Dr. Deftos received his undergraduate degree from Brown University and his MD from the University of Vermont. He trained at Columbia-Presbyterian Medical Center, the National Institutes of Health, and the Massachusetts General Hospital, and he was on the staff of the latter and the faculty at Harvard Medical School before moving to California. He is a board-certified internist and endocrinologist with a special interest in calcium and skeletal metabolism that spans over 25 years and is represented by over 700 publications, including many textbook chapters. In addition to his medical degree, Dr. Deftos has a JD and writes about legal issues in medicine, as illustrated by the genetic testing issues in Chapter 6 of this book.

This book is designed for a wide audience and is based on Dr. Deftos' experience as a teacher of medical and graduate students, housestaff, postgraduate fellows, staff and faculty. The book will be of interest to all students, interns, residents and postgraduate students with an interest in skeletal diseases; to faculty with such interests; to the practicing primary-care physician who sees patients with common diseases such as osteoporosis; and as a succinct review for the specialist who sees patients with calcium and skeletal diseases.

Preface

This handbook is designed as a basic primer for individuals interested in calcium and skeletal metabolism and their disorders. It is intended to have wide appeal, ranging from students to practitioners. The text is based on the notes and slides that I have developed for teaching medical and other postgraduate students at the University of California, San Diego and for corresponding lectures to faculty and staff. In developing these teaching aids, I have unabashedly borrowed from my colleagues, usually with their express consent, but always with at least implied consent. I have most recently relied on the subsequently listed publications. Additional suggested readings are recommended at the end of each chapter. The compact size of the handbook gives it a portability and economy that textbooks do not have, and the publication process allows timely discussion. On the other side of the coin, however, the size does not allow the detailed exposition contained in the many excellent texts in this area, some which are mentioned below. I welcome my colleagues' comments on the handbook. I can be reached at ljdeftos@ucsd.edu.

SUGGESTED GENERAL READINGS

Bramley I, ed. Bailliére's *Clinical Endocrinology and Metabolism: International Practice and Research. Metabolic Bone Disease.* Vol 11. London, UK: Bailliére Tindall; 1997.

Bilezikian JP, Marcus R, Levine MA, eds. *The Parathyroids: Basic and Clinical Concepts.* New York, NY: Raven Press; 1994.

Bilezikian JP, Raisz LG, Rodan GA, eds. *Principles of Bone Biology.* San Diego, Calif: Academic Press, Inc; 1996.

Coe FL, Favus MJ, eds. *Disorders of Bone and Mineral Metabolism*. New York, NY: Raven Press; 1992.

Deftos LJ, Glowacki J. Mechanisms of bone metabolism. In: Kem DC, Frohlich E, eds. *Pathophysiology*. 3rd ed. Philadelphia, Pa: JB Lippincott Co; 1984:445-468.

Favus MJ, ed. *Primer on Metabolic Bone Diseases and Disorders of Mineral Metabolism.* 3rd ed. Philadelphia, Pa: Lippincott-Raven; 1996.

Greenspan FS, Strewler GJ, eds. *Basic and Clinical Endocrinology*. 5th ed. Norwalk, Conn: Appleton and Lange; 1997. (Especially the chapters by Strewler GJ, *Mineral Metabolism & Metabolic Bone Disease*, Chapter 8:263-317; and Deftos LJ, Nolan JJ, *Syndromes Involving Multiple Endocrine Glands*, Chapter 24:753-770.)

Marcus R, Feldman D, Kelsey J, eds. *Osteoporosis*. San Diego, Calif: Academic Press; 1996.

Mundy GR. *Calcium Homeostasis: Hypercalcemia nd Hypocalcemia*. 2nd ed. London, UK: Martin Dunitz Ltd; 1990.

Rosen CJ, ed. *Osteoporosis: Diagnostic and Therapeutic Principals*. Totowa, NJ: Humana Press; 1996.

Vassallo J, ed; Lovette HS, production-ed. *Endocrinology and Metabolism Clinics of North America*. Philadelphia, Pa: W.B. Saunders Company; June, 1998. Watts NB, ed. Osteoporosis; Vol. 27, No. 2.

Wilson JD, Foster DW, Kronenberg HM, Larsen PR, eds. *Williams Textbook of Endocrinology*. 9th ed. Philadelphia, Pa: W.B. Saunders Co; 1998. (Especially the chapters by Bringhurst FR, Demay MB, Kronenberg HM, *Hormones and Disorders of Mineral Metabolism*, Chapter 24:1155-1210; Raisz LG, Kream BE, Lorenzo JA, *Metabolic Bone Disease*, Chapter 25:1211-1240; and Pak CYC, *Kidney Stones*, Chapter 26:1241-1258.)

PART I

Regulation of Skeletal and Calcium Metabolism

1

Skeletal Metabolism

1

Introduction

The skeleton has two unique and important functions:

- Metabolic
- Mechanical.

The metabolic function of bone is to provide a reservoir for ions, such as:

- Calcium
- Phosphorus
- Magnesium.

The mechanical structure of bone:

- Protects vital organs
- Supports locomotion by its muscle attachments
- Encases bone marrow.

The latter provides proximity between bone and the bone precursor cells of the marrow.

Bone minerals can be mobilized to maintain systemic mineral homeostasis. Bone is also a depository for certain cytokines and growth factors that can be released upon its resorption and exert their effects locally and systemically. The metabolic function of bone prevails over its structural function in that calcium is removed from and replaced in bone to serve systemic homeostatic needs irrespective of loss of skeletal structural integrity.

Structure of Bone

There are two major types of bone:

- Cortical (also referred to as compact)
- Trabecular (also referred to as cancellous or spongy).

Most bone is cortical bone and provides for the mechanical function of the skeleton. Cortical bone is composed of dense layers of mineralized collagen arranged concentrically or in parallel perpendicular planes and is found primarily in the appendicular skeleton — the outer casing of the long bones of the arms and legs and of the vertebra. There is a gradual transition from cortical bone at the mid-shaft of long bones to trabecular bone at their heads. Cortical bone comprises approximately 80% of the skeleton and trabecular bone 20%; the former provides about 75% of bone mass, the latter about 75% of bone surface. Trabecular bone is spongy in structure and is found primarily in the axial skeleton and the vertebral bodies. However, the surface area of cortical bone is only one fifth that of trabecular bone. Thus, trabecular bone is metabolically more active than cortical bone, with an annual turnover (remodeling) of approximately 20% to 30% for the former and 3% to 10% for the latter. A given skeletal site in the adult is remodeled approximately every 3 years. Bone mass is acquired up to the fourth to fifth decade, with a rapid phase during adolescent growth. Most of peak bone mass is genetically determined (Figure 1.1). Women have approximately 30% less peak bone mass than men and experience an accelerated loss after the menopause.

Bone consists of a mineral phase and an organic phase. The major component of the mineral phase is hydroxyapatite crystal and the major component of the organic phase is type I collagen which with other proteins comprises the matrix of bone (Table 1.1).

FIGURE 1.1 — PEAK BONE MASS

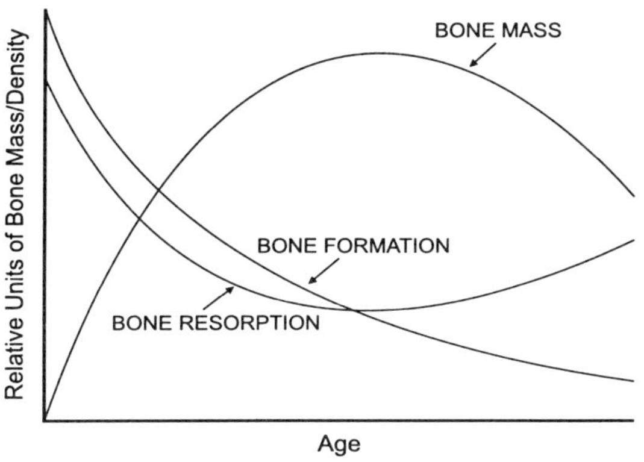

Schematic representation in relative units of normal skeletal development, demonstrating changes in bone resorption and formation. The crossover of formation/resorption occurs during the fourth decade. In osteoporosis, there is an accelerated loss of bone because of increased resorption and decreased formation. See also Figure 10.1.

■ Bone Matrix

The major organic component of bone is type I collagen, comprising about 90% of the organic phase of bone. There are also many noncollagenous organic components of bone (Table 1.2). Several cytokines (Table 1.3) and growth factors (Table 1.4) have profound skeletal effects, and some are present in bone matrix and can be released to exert their biological effects, notably transforming growth factor-beta (TGF-β). The status of the organic matrix of bone can be assessed in blood and urine by measuring its breakdown products, which include hydroxyproline and collagen cross-links.

■ Bone Mineral

Bone mineral is present in two forms in the skeleton. Hydroxyapatite crystals, represented by the for-

TABLE 1.1 — STRUCTURE OF BONE
Cellular
• Osteocytes • Osteoblasts • Osteoclasts • Lining cells
Noncellular
• Mineral – Hydroxyapatite: $Ca_{10}(PO4)_6(OH)_2$ • Matrix – Collagen: type I – Glycosaminoglycan-containing proteins – Glycoproteins – Gamma carboxyglutamic acid (GLA)-containing proteins – Other proteins

mula $Ca_{10}(PO4)_6(OH)_2$, are the major forms and occur in mature bone. Amorphous calcium phosphate comprises the remainder; it occurs in areas of active bone formation and matures through several intermediate stages to hydroxyapatite. Mineralization occurs in the molecular "hole" that is created between the ends of collagen molecules as they assume their staggered array. Matrix vesicles released from chondrocytes may also participate in mineralization by releasing their alkaline phosphatase (AP) and providing a nidus. Mineralization of osteoid depends on the ambient concentrations of calcium, phosphate, and other ions. Mineralization can occur if:

- There is a functionally adequate local concentration of these ions
- Nucleators are present to promote crystallization and consequent mineralization
- Local inhibitors of mineralization are removed.

TABLE 1.2 — BONE PROTEINS
Collagen (Type I)
Glycosaminoglycan-Containing Proteins • Aggrecan • Versican • Small proteoglycans – Decorin (PG-II, PG-40) – Biglycan (PG-I, PG-SI) – Fibromodulin • Heparan sulfate proteoglycans • Other small proteoglycans • Hyaluronan
Glycoproteins • Alkaline phosphatase • Osteonectin (SPARC, BM-40) • Tetranectin • Arginine, glycine, aspartic acid (RGD)-containing glycoproteins – Thrombospondin – Fibronectin – Vitronectin – Osteopontin (formerly BSP-1, SPP, pp66) – Bone sialoprotein – Bone acidic glycoprotein-75 (BAG-75) – Fibrillin – Other RGD-containing proteins
Gamma Carboxyglutamic Acid (GLA)-Containing Proteins • Osteocalcin (BGP) • Matrix GLA protein (MGP) • Protein S
Other Proteins • Proteolipids • Enzymes and inhibitors – Metalloproteinases – Plasminogen activator and plasminogen activator inhibitor – Matrix phosphoprotein kinases • Bone morphogenetic proteins • Growth factors • Serum proteins (α2-HS-glycoprotein, albumin, etc)

TABLE 1.3 — CYTOKINES THAT INCREASE BONE RESORPTION		
Cytokine	**Increase Osteoclast Formation**	**Increase Osteoclast Activity**
IL-1	+	+
IL-6	+	–
IL-11	+	–
CSFs	+	–
Abbreviations: IL, interleukin; CSF, colony-stimulating factor.		

TABLE 1.4 — BONE GROWTH FACTORS
• Transforming growth factor-beta (TGF-β) family of polypeptides • Platelet-derived growth factors • Fibroblast growth factors • Insulin-like growth factors • Insulin-like growth factor binding proteins

All these mechanisms are likely to participate in both normal and pathological mineralization.

■ Bone Cells

There are four types of bone cells (Table 1.1):

- Osteoblasts
- Osteocytes
- Osteoclasts
- Lining cells.

Monocytes, macrophages, and mast cells may also mediate certain aspects of skeletal metabolism. Marrow cells contribute to the population of bone cells.

The Osteoblast

The osteoblast forms bone. It develops from mesenchymal precursor cells under the influence of hormones, cytokines, and growth factors (Table 1.5). Osteoblasts express receptors to many bone-active agents such as parathyroid hormone (PTH), parathyroid hormone-related protein (PTHrP), vitamin D metabolites, gonadal and adrenal steroids, and certain cytokines and lymphokines. Response by these receptors allows the osteoblast to initiate osteoclastic bone resorption by secreting collagenase and other metalloproteinases and plasminogen activator to pro-

TABLE 1.5 — OSTEOBLAST REGULATION*
Stimulation • Leukemia inhibitory factor (LIF) • Interleukin 10 (IL-10) • Platelet-derived growth factor (PDGF) • Fibroblast growth factor (FGF) • Epidermal growth factor (EGF) • Parathyroid hormone/parathyroid hormone-related protein (PTH/PTHrP) • Insulin • Insulin-like growth factor-I (IGF-I) • Gonadal steroids • Oncostatin M
Inhibition • Glucocorticoids • Bax/bcl-2 • Tumor necrosis factor (TNF)
Biphasic Regulation • 1,25-Dihydroxyvitamin D (1,25-D) • Bone morphogenetic proteins (BMPs) 1, 2, 4, 6 • Transforming growth factor-β (TGF-β)
* These agents act on osteoblasts by regulating their genesis, growth and proliferation, differentiation, and/or bone formation function (see Figure 1.2)

vide matrix access to the osteoclast. The major product of osteoblasts is type 1 collagen, which along with other proteins (Table 1.2) forms the organic osteoid matrix that is mineralized by hydroxyapatite. Osteoblasts also produce other proteins that participate in the mineralization process. Osteoblast activity can thus be assessed by measuring in serum some of its biosynthetic products such as AP and osteocalcin and also certain collagen precursor peptides. Osteoblasts form a layer of cells on bone surfaces. Mineralization occurs when there is a functionally adequate ambient supply of:

- Calcium
- Phosphorus
- Other hydroxyapatite elements.

Osteoblast precursors (mesenchyme) can also develop into:

- Fibroblasts
- Chondrocytes
- Adipocytes
- Myocytes.

The Osteocyte

Osteocytes are osteoblasts that become encased in bone during its formation and mineralization and reside in the resulting lacuna. While their synthetic activity decreases, the cells develop processes that communicate through canaliculi with other osteocytes, osteoblasts, and the vasculature (Haversian canal). Osteocytes present acres of cellular syncytium that permits translocation of bone mineral during times of metabolic activity and can provide minute-to-minute exchanges of minerals from bone matrix, although this function is disputed. Osteocytes are also the likely transducers through their canaliculi of mechanic forces on bone and, in concert with other bone cells, mediate Wolf's law — the complex remodeling response to mechanical stimuli of the skeleton that causes appropriate changes in formation and resorption in response to skeletal loading.

The Osteoclast

The osteoclast resorbs bone. It is a terminally-differentiated, large, multinucleated giant cell that arises from hematopoietic marrow precursors under the influences of hormones, growth factors, and cytokines (Table 1.6). Osteoclast precursors arise from the colony-forming unit (CFU) of granulocytes and macrophages under the regulatory interactions of stem-cell factor (SCF), interleukin 1 (IL-1), IL-6, IL-11, TGF-β, TNF, and several CSFs, notably granulocyte-macrophage (GM)-CSF and CSF-1 (aka macrophage CSF-MCSF). Under this complex regulation, mononuclear osteoclast precursors fuse into mature, terminally-differentiated, multinucleated osteoclasts.

The osteoclast resorbs bone by attachment with a ruffled border through adhesion molecules and by secretion of hydrogen and chloride ions that dissolve mineral and lytic proteases, notably lysosomal proteases active at low pH and metalloproteinases and cysteine proteinases that dissolve matrix. The "pit" of resorption is called a Howship lacuna. In contrast to the receptor-rich osteoblast, the mature osteoclast has few receptors, but it robustly expresses the receptor for CT. The proto-oncogenes *c-fos* and *c-src* are important for the microarchitectural and signal transduction changes that allow osteoclasts to resorb bone. An osteoclast remains active for about 2 weeks, then undergoes apoptosis. Figure 1.2 summarizes osteoblast and osteoclast lineage.

Bone-Lining Cells and Other Cells

Bone-lining cells are flat, elongated cells that cover inactive bone surfaces. Their function is unknown, but they may be osteoblast precursors. Mast

TABLE 1.6 — OSTEOCLAST REGULATION*
Stimulation • Parathyroid hormone/parathyroid hormone-related protein (PTH/PTHrP) • 1,25-Dihydroxyvitamin D (1,25-D) • Thyroid hormones • Interleukin 1α, 3, 6, 11 (IL-1α, -3, -6, -11) • Macrophage colony-stimulating factor (M-CSF, CSF-1) • Granulocyte-macrophage colony-stimulating factor (GM-CSF) • Tumor necrosis factor-α (TNF-α) • Osteoprotegerin (OPG) ligand • Transforming growth factor-α (TGF-α) • Epidermal growth factor (EGF) • Leukotrienes • Vitamin A • Stem-cell factor (SCF) • Basic fibroblast growth factor (bFGF) • Erythropoetin (EPO) • c-FOS and c-SRC • Caspaces and adhesion molecules • Insulin-like growth factor-I (IGF-I) • Bone morphogenetic protein (BMP) 2, 4 • 5-Lipoxygenase (5-LO) metabolites • Chaperonin 60
Inhibition • Calcitonin (CT) • Gonadal steroids (including 17-beta estradiol) • Interferon-γ (INF-γ) • Interleukin-4 (IL-4) • Nitric oxide (NO)
Biphasic Regulation • Prostaglandin E_2 (PGE_2) • Transforming growth factor-β (TGF-β) • Glucocorticoids
* These agents act on osteoclasts by regulating their genesis, growth and proliferation, differentiation, fusion, attachment, polarization, and/or bone resorption function (see Figure 1.2)

FIGURE 1.2 — SCHEMATIC REPRESENTATION OF OSTEOCLAST AND OSTEOBLAST LINEAGES

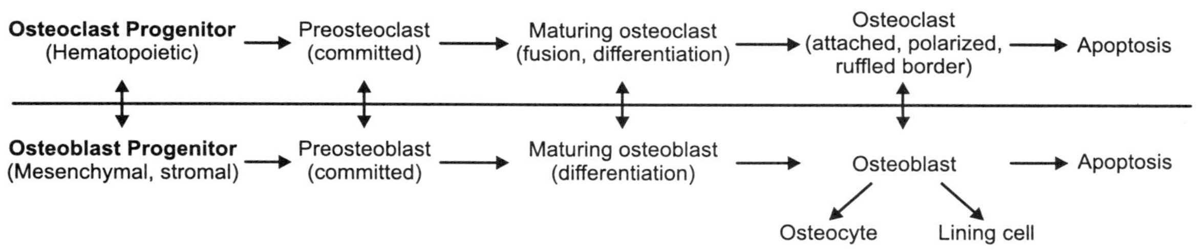

Schematic representation of the osteoclast (top) and osteoblast (bottom) lineages. The two lineages are distinct, but there is regulatory interaction among the cells (vertical arrows). Osteoclasts originate from a hematopoietic stem cell that can also differentiate into a macrophage, granulocyte, erythrocyte, megakaryocyte, mast cell, B-cell, or T-cell. Osteoblasts originate from a mesenchymal stem cell that can also differentiate into a chondrocyte, myocyte, fibroblast, or adipocyte. The terminology for these lineages is still evolving and is herein [over] simplified. Many intermediate steps and regulatory factors are involved in lineage development. (See Tables 1.5 and 1.6.)

cells can be seen at sites of bone resorption and may participate in this process. Cells of the immune system play a key role in bone metabolism, especially resorption.

Bone Metabolism

■ Growth, Modeling and Remodeling

Growth, modeling and remodeling are important processes that allow the skeleton to play its many important roles. Bone grows and models under the influence of metabolic, mechanical, and gravitational forces during growth through adolescence, changing its size and shape in the process. Bone growth continues until approximately age 18. Bone mass continues to increase until the fourth decade (Figure 1.1). Most of the skeleton develops from cartilage by a process called endochondral ossification, whereby cartilage is resorbed and replaced by bone. However, the vault of the skull, clavicle, facial bones, mandible and maxilla develop from mesenchyme by a process called intramembranous ossification. Bone renews itself by remodeling, a cycle in which old bone is first resorbed and new bone is then formed to replace it. Both cortical bone and trabecular bone remodel, but the latter is more metabolically active.

Bone remodeling can be divided into four stages:

- Activation (osteoblasts)
- Resorption (osteoclasts)
- Reversal (lining cells)
- Formation (osteoblasts).

Remodeling serves to repair skeletal microdamage and to improve skeletal strength in response to mechanical forces.

Cortical bone is remodeled by "cutting cones" of osteoclasts that tunnel through it. Trailing osteoblasts lining the resulting tunnels deposit new bone on their walls. The osteoblasts produce bone matrix which becomes mineralized. The tunnels gradually narrow and become the Haversian canals that nourish the resident osteocytes through their blood vessels. The unit of new bone thus formed is called an osteon.

1

Trabecular bone remodels on its surface. Osteoclasts first resorb a "pit" of bone, and osteoblasts ultimately fill it in with new bone. In between the resorption and formation sites, a cement line is deposited. The sequence of resorption and formation is called a basic multicellular unit (BMU) of bone turnover. The life-span of a BMU is estimated at 2 to 8 months. The remodeling cycle in cortical and trabecular bone takes approximately 150 days, mostly by osteoblastic bone formation. But because of its large surface-to-volume ratio, the turnover rate of trabecular bone is more rapid in both health and disease.

Most remodeling occurs in trabecular bone and on the endosteal surfaces of cortical bone, with little periosteal remodeling. However in diseases like hyperparathyroidism, subperiosteal resorption is activated. With aging, periosteal remodeling and expansion seems to compensate (mechanically) for bone loss at other sites.

Osteoclasts and osteoblasts communicate with each other during remodeling in a process that is referred to as coupling and that is mediated by local regulatory signals that have not been as yet identified. Coupling assures a balance of bone formation and bone resorption in the adult skeleton. Figure 1.3 summarizes the bone-remodeling process in the overall scheme of calcium and skeletal metabolism.

■ Regulatory Factors

A variety of hormones (Table 1.7), bone morphogenetic proteins (Table 1.8), cytokines (Table 1.3), and growth factors (Table 1.4) regulate skeletal metabo-

FIGURE 1.3 — SCHEMATIC REPRESENTATION OF CALCIUM AND SKELETAL METABOLISM

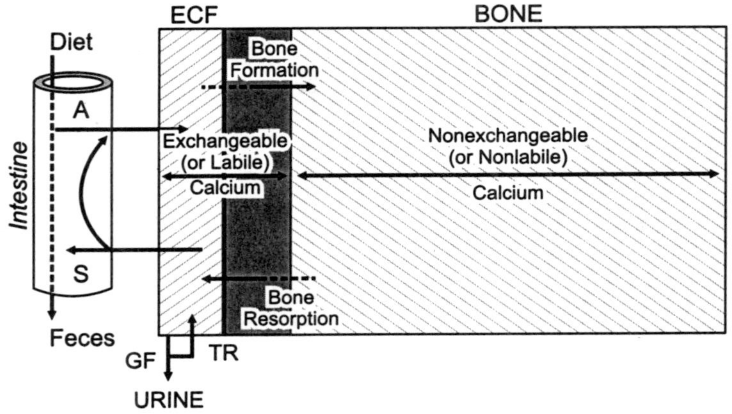

Abbreviations: A, absorption; S, secretion; ECF, extracellular fluid; GF, glomerular filtration; TR, tubular reabsorption.

The dark vertical line between bone and ECF represents bone surface and bone-lining cells. Shaded area represents labile skeletal calcium. The various calcium compartments are not to scale. See text for discussion.

TABLE 1.7 — HORMONES THAT REGULATE BONE METABOLISM
Increase Bone Resorption • Parathyroid hormone/parathyroid hormone-related protein (PTH/PTHrP) • Glucocorticoids • Thyroid hormones • High dose vitamin Ds
Decrease Bone Resorption • Calcitonin • Gonadal steroids
Increase Bone Formation • Growth hormone • Vitamin D metabolites • Gonadal steroids • Insulin • Low-dose PTH/PTHrP
Decrease Bone Formation • Glucocorticoids

TABLE 1.8 — BONE MORPHOGENETIC PROTEINS
• Bone morphogenetic protein 2 (BMP-2) • BMP-3 • BMP-4 • BMP-5 • BMP-6 • Osteogenic protein-1 (OP-1, aka BMP-7) • Osteogenic protein-2 (OP-2, aka BMP-8) • BMP-9 • Growth-differentiation factors (GDF-5 to GDF-7)

1

lism. Hormonal effects are discussed in detail in Chapter 2, *Calcium Metabolism.*

Equally important are local regulatory factors that are produced by bone cells and adjacent hematopoietic cells and stored in bone matrix. Those that increase bone resorption are:

- Il-1, -6, and -11 (Table 1.3)
- TNFs
- CSFs
- TGF-α
- Epidermal growth factor.

Those that increase bone formation (Table 1.4) are:

- Insulin-like growth factors I and II
- TGF-β
- Bone morphogenetic proteins (Table 1.8)
- Fibroblast growth factors 1 and 2
- Platelet-derived growth factor.

Those that have bimodal effects are:

- Prostaglandins of the E series
- IL-4.

■ Bone Markers

During growth, modeling, and remodeling, the metabolic activity of bone cells on bone matrix releases products into blood that can be measured to assess skeletal metabolism. These bone markers can be measured in blood and urine to assess the overall status of bone metabolism and to respectively assess bone formation (Table 1.9) and bone resorption (Table 1.10). Although their clinical use is controversial, especially for diagnostic purposes, recent studies indicate that these markers may be used to evaluate therapy and predict fracture risk in osteoporosis (see Chapter 10, *Osteoporosis*).

Many bone marker tests are now commercially available. They reflect the activity of osteoblasts, os-

TABLE 1.9 — BIOCHEMICAL MARKERS OF BONE FORMATION

	Common Names/ Abbreviations
Total alkaline phosphatase	AP
Bone-specific alkaline phosphatase	BAP
Osteocalcin	OC, BGP
C- and N-terminal propeptides of type I procollagen	PICP, PINP

TABLE 1.10 — BIOCHEMICAL MARKERS OF BONE RESORPTION

	Common Names/ Abbreviations
N- and C-terminal cross-linked telopeptides of type I collagen	NTX, CTX
Pyridinoline (total, free)	PYD
Deoxypyridinoline (total, free)	DPD
Hydroxyproline and hydroxylysine glycosides	OH, OL
Tartrate-resistant acid phosphatase	TRAP

teoclasts, and the skeleton itself. Resorption markers change within a few weeks of skeletal perturbation, formation markers within a few months. Formation markers are generally measured in serum and resorption markers in urine, but serum markers for the latter are being introduced. Figure 1.4 is a schematic representation of their release into blood (and urine). Their clinical applications are discussed in Part III of this text, especially Chapter 10, *Osteoporosis*. In general, bone markers can be used to evaluate therapy but not for diagnosis. Most importantly, they provide

FIGURE 1.4 — SCHEMATIC REPRESENTATION OF THE CELLULAR AND SKELETAL SOURCES OF SERUM AND/OR URINARY MARKERS OF BONE FORMATION AND BONE RESORPTION

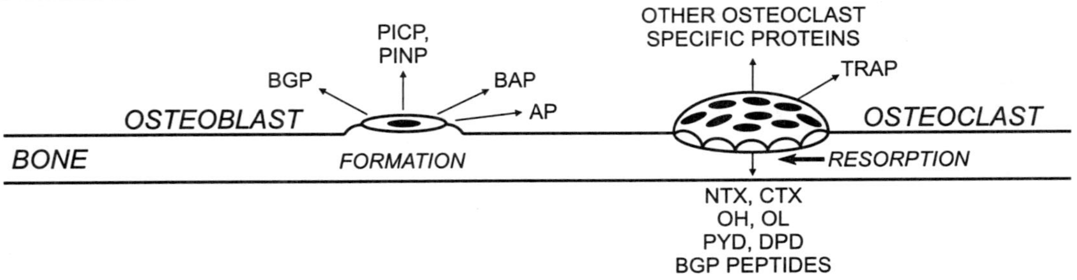

Abbreviations: BGP, bone gamma carboxyglutamic acid (GLA) protein (osteocalcin); PICP, C-terminal propeptide of type I procollagen; PINP, N-terminal propeptide of type I procollagen; BAP, bone-specific alkaline phosphatase; AP, alkaline phosphatase; TRAP, tartrate-resistant acid phosphatase; NTX, N-terminal cross-linked telopeptide of type I collagen; CTX, C-terminal cross-linked telopeptide of type I collagen; OH, hydroxyproline glycoside; OL, hydroxylysine glycoside; PYD, pyridinoline (total, free); DPD, deoxypyridinoline (total, free).

See Table 1.9 and text for discussion.

valuable research tools. The markers with most clinical evaluation are AP and osteocalcin for bone formation and pyridinolines and collagen cross-links for bone resorption.

1

Formation Markers

The formation markers AP, bone-specific alkaline phosphatase (BAP) and osteocalcin (OC), directly reflect osteoblast activity. Routine AP testing measures total AP enzymatic activity, so it can reflect other organs (eg, liver) as well as bone. Nevertheless, it remains a most widely used test, but its clinical value is greatest in Paget's disease (Chapter 8, *Paget's Disease of Bone* [*Osteitis Deformans*]). Specificity can be improved by chromatographic separation of the different species of AP and, less specifically, by heat inactivation of the bone form. BAP measures the bone form of AP immunochemically, but there is some assay cross-reactivity (~ 15%) with liver AP.

There are many available immunoassays for OC. Depending on format and antibody specificity, they measure different components of the molecule. The major circulating product of intact OC is the 1-43 peptide, and smaller peptides may reflect collagen degradation.

C-terminal propeptides of type I collagen (PICP) and N-terminal propeptides of type I procollagen (PINP) reflect osteoblast activity, but also some collagen degradation, especially PINP. However, these measurements, especially the former, do not appear to reflect changes in bone metabolism as well and are not as widely used.

Resorption Markers

Cross-links of collagen peptides, both amino-terminal (NTX) and carboxy-terminal (CTX), especially the former, are widely used resorption markers that reflect collagen breakdown. Tartrate-resistant acid phosphatase (TRAP) reflects osteoclast activity but must be distinguished from enzyme activity derived from other tissue sources, notably prostate, pancreas, and blood cells. This is usually accomplished by electrophoresis and by the resistance of its activity to the inhibitor tartrate. However, even the tartrate resistant form of the enzyme may be present in other cells. Immunoassays have been developed for TRAP, but there have been only limited studies of them in patients with skeletal diseases. Nevertheless, the results are consistent with the hypothesis that serum TRAP is a marker for cellular osteoclastic bone resorption. The value of these assays for clinical application has not yet been established.

Pyridinoline cross-links are released into the circulation during bone resorption and are excreted in urine as both free pyridinoline (PYD) and deoxypyridinoline (DPD) and pyridinolines bound to bone specific C- or N-termini of type I collagen. DPD is more specific for bone collagen than PYD. The most sensitive and accurate methodology for measuring urinary cross-links of the pyridinolines is high performance liquid chromatography (HPCL). However, only a few laboratories have this complex and laborious methodology, and immunochemically-based assays are used by most commercial laboratories.

The measurement of free DPD and PYD represent their respective pyridinium moieties, whereas their total measurements include their cross-linked collagen peptides. There have been recent advances in the development of better methods to measure pyridinolines and their cross-links both in urine and serum. These tests rely on antibodies that have varying specificity to the peptides and their intermolecular cross-link domain and the pyridinolines. These markers are usually measured in urine, but similarly developed serum tests are becoming available. Furthermore, 24-hour urine collections are being replaced by spot collections for some assays.

Urinary calcium and hydroxyproline have also been used to assess bone resorption. Despite improvements in methodology for the latter, their use has been supplanted by the more specific methods.

Table 1.11 summarizes the potential and/or actual application of bone markers in bone disease.

TABLE 1.11 — APPLICATION AND RELATIVE VALUE OF BONE MARKERS IN SKELETAL DISEASE
Research Tools (+++)
Therapy (++) • Evaluate effectiveness of therapy • Monitor compliance to therapy • Adjust therapy to effective doses
Prognosis (+) • Assess fracture risk • Predict rates of bone loss and gain • Identify patients for appropriate therapy
Diagnosis (+/–)

■ Skeletal Assessment

The skeleton can be assessed by biopsy, x-rays, and other imaging procedures (see Table 10.7). Bone biopsy is seldom used for routine clinical purposes. However, it remains a useful procedure that can provide information about bone formation and bone resorption. As detailed in Chapter 10, *Osteoporosis*, newer imaging procedures allow the assessment of bone density, mass per unit volume. True bone density changes little with age, but apparent bone density, mass divided by volume, declines with age in the adult and correlates with bone strength. Clinically-used densitometric procedures only approximate bone density in that they measure mass as a function of estimated volume. Nevertheless, such measurements, limitations notwithstanding, are important clinical tools.

Summary

The skeleton is the reservoir of calcium for many physiological functions, and it serves a similar but not so unique role for phosphorus and magnesium. Skeletal calcium is regulated through the calcium pathways of the gastrointestinal (GI) tract and the kidney, and this regulation is mediated in bone by the osteoblast, the bone-forming cell, and the osteoclast, the bone-resorbing cell. Calcium reaches the skeleton by being absorbed from the diet in the GI tract. Unabsorbed calcium passes into the feces, which also contains the small amount of calcium secreted into the GI tract. Dietary calcium is then absorbed into the extracellular fluid (ECF) space and becomes incorporated into the skeleton through the process of mineralization of osteoid. ECF calcium is also filtered by the kidney, where most of it is reabsorbed.

Collagen (type I), the primary organic component of osteoid, is produced by the osteoblast, along with most of the other proteins that comprise this organic phase of bone. Mineralization proceeds through intermediary steps to form hydroxyapatite, the crystal structure of the other phase of bone – the nonorganic, mineral phase. The end result is a highly organized amalgam of protein, primarily collagen, and mineral, primarily hydroxyapatite, that has sufficient structural integrity to serve the mechanical functions of the skeleton. Upon completion of this process, the osteoblast becomes encased in bone and become an osteocyte.

In the adult skeleton, this process of bone formation is balanced by the process of bone resorption. Bone resorption is mediated by the osteoclast, a large, multinucleated cell that is molecularly equipped to dissolve both the mineral and organic phases of bone. After completing its function, the terminally-differentiated osteoclast undergoes apoptosis. The processes of osteoblast-mediated bone formation and osteoclast-mediated bone resorption are "coupled." This bone metabolism can be assessed by measurement of bone markers. Approximately 20% of adult bone surface is undergoing remodeling at any time. Osteoblasts, osteoclasts, and perhaps osteocytes are regulated by many hormones, cytokines, and growth factors. The homeostatic end-point of skeletal metabolism is to provide the appropriate amount of ambient calcium for the many biological functions that this ion serves.

SUGGESTED READING

Albright F. Postmenopausal osteoporosis. *JAMA*. 1941;116:2465-2474.

Amano H, Yamada S, Felix R. Colony-stimulating factor-1 stimulates the fusion process in osteoclasts. *J Bone Miner Res*. 1998;13:846-853.

Bienkowski RS, Cowan MJ, McDonald JA, Crystal RG. Degradation of newly synthesized collagen. *J Biol Chem*. 1978;253:4356-4363.

Bikle DD. Biochemical markers in the assessment of bone disease. *Am J Med*. 1997;103:427-436.

Brown EM, Pollak M, Hebert SC. The extracellular calcium-sensing receptor: its role in health and disease. *Annu Rev Med*. 1998;49:15-29.

Cohen-Solal ME, Boitte F, Bernard-Poenaru O, et al. Increased bone resorbing activity of peripheral monocyte culture supernatants in elderly women. *J Clin Endocrinol Metab*. 1998;83:1687-1690.

Deftos LJ. Immunoassay for human calcitonin. I. Method. *Metabolism*. 1971;20:1122-1128.

Delmas PD. Biochemical markers of bone turnover for the clinical assessment of metabolic bone disease. *Endocrinol Metab Clin North Am.* 1990;19:1-18.

Delmas PD, Schlemmer A, Gineyts E, Riis B, Christiansen C. Urinary excretion of pyridinoline crosslinks correlates with bone turnover measured on iliac crest biopsy in patients with vertebral osteoporosis. *J Bone Miner Res.* 1991;6:639-644.

Dong, Y, Qiu QQ, Debear J, Lathrop WF, Bertolini DR, Tamburini PP. 17β-Hydroxysteroid dehydrogenases in human bone cells. *J Bone Miner Res.* 1998;13:1539-1546.

Everts V, Delaissé J-M, Korper W, Beertsen W. Cysteine proteinases and matrix metalloproteinases play distinct roles in the subosteoclastic resorption zone. *J Bone Miner Res.* 1998;13:1420-1430.

Eyre D. New biomarkers of bone resorption. *J Clin Endocrinol Metab.* 1992;74:470A-470C. Editorial.

Eyre DR, Koob TJ, Van Ness KP. Quantitation of hydroxypyridinium crosslinks in collagen by high-performance liquid chromatography. *Anal Biochem.* 1984;137:380-388.

Fox SW, Chow JW. Nitric oxide synthase expression in bone cells. *Bone.* 1998;23:1-6.

Garnero P, Shih WJ, Gineyts E, Karpf DB, Delmas PD. Comparison of new biochemical markers of bone turnover in late postmenopausal osteoporotic women in response to alendronate treatment. *J Clin Endocrinol Metab.* 1994;79:1693-1700.

Gertz BJ, Shao P, Hanson DA, et al. Monitoring bone resorption in early postmenopausal women by an immunoassay for cross-linked collagen peptides in urine. *J Bone Miner Res.* 1994;9:135-142.

Guglielmi G, Selby K, Blunt BA, et al. Magnetic resonance imaging of the calcaneus: preliminary assessment of trabecular bone–dependent regional variations in marrow relaxation time compared with dual x-ray absorptiometry. *Acad Radiol.* 1996;3:336-343.

Hamilton BA, MePhee JL, Hawrylak K, Stinson RA. Alkaline phosphatase releasing activity in human tissues. *Clin Chim Acta.* 1990;186:249-254.

Harada H, Kukita T, Kukita A, Iwamoto Y, Iijima T. Involvement of lymphocyte function-associated antigen-1 and intercellular adhesion molecule-1 in osteoclastogenesis: a possible role in direct interaction between osteoclast precursors. *Endocrinology*. 1998;139:3967-3975.

Harvey RD, McHardy KC, Reid IW, et al. Measurement of bone collagen degradation in hyperthyroidism and during thyroxine replacement therapy using pyridinium cross-links as specific urinary markers. *J Clin Endocrinol Metab*. 1991;72:1189-1194.

Heany RP. Bone mass, bone loss, and osteoporosis prophylaxes. *Ann Intern Med*. 1979;103:313-314.

Holick MF. Perspective on the impact of weightlessness on calcium and bone metabolism. *Bone*. 1998;22(suppl 5):105S-111S.

Jilka RL. Cytokines, bone remodeling, and estrogen deficiency: a 1998 update. *Bone*. 1998;23:75-81.

Jilka RL, Weinstein RS, Bellido T, Parfitt AM, Manolagas SC. Osteoblast programmed cell death (apoptosis): modulation by growth factors and cytokines. *J Bone Miner Res*. 1998;13:793-802.

Kameda T, Mano H, Yamada Y, et al. Calcium-sensing receptor in mature osteoclasts, which are bone resorbing cells. *Biochem Biophys Res Commun*. 1998;245:419-422. Abstract.

Kawakami A, Nakashima T, Tsuboi M, et al. Insulin-like growth factor I stimulates proliferation and fas-mediated apoptosis of human osteoblasts. *Biochem Biophys Res Commun*. 1998;247:46-51.

Kivirikko KI, Laitinen O, Prockop DJ. Modifications of a specific assay for hydroxyproline in urine. *Anal Briochem*. 1967;19:249-255.

Kronenberg HM. Parathyroid hormone and osteocalcin—when friends become strangers. *Endocrinology*. 1997;138:3083-3084. Editorial.

Lacey DL, Timms E, Tan HL, et al. Osteoprotegerin ligand is a cytokine that regulates osteoclast differentiation and activation. *Cell*. 1998;93:165-176.

Lader CS, Flanagan AM. Prostaglandin E2, interleukin 1α, and tumor necrosis factor-α increase human osteoclast formation and bone resorption *in vitro*. *Endocrinology*. 1998;139:3157-3164.

Lian JB Gundberg CM. Osteocalcin. Biochemical considerations and clinical applications. *Clin Orthop*. 1988;226:267-291.

Masters PW, Jones RG, Purves DA, Cooper EH, Cooney JM. Commercial assays for serum osteocalcin give clinically discordant results. *Clin Chem*. 1994;40:358-363.

Mizuno A, Amizuka N, Irie K, et al. Severe osteoporosis in mice lacking osteoclastogenesis inhibitory factor/osteoprotegerin. *Biochem Biophys Res Commun*. 1998;247:610-615.

Okahashi N, Koide M, Jimi E, Suda T, Nishihara T. Caspases (interleukin-1 beta-converting enzyme family proteases) are involved in the regulation of the survival of osteoclasts. *Bone*. 1998;23:33-41.

Parfitt AM. A structural approach to renal bone disease. *J Bone Min Res*. 1998;13:1213-1220.

Poser JW, Esch FS, Ling NC, Price PA. Isolation and sequence of the vitamin K-dependent protein from human bone. Undercarboxylation of the first glutamic acid residue. *J Biol Chem*. 1980;225:8685-8691.

Price PA, Parthemore JG, Deftos LJ. New biochemical marker for bone metabolism. Measurement by radioimmunoassay of bone GLA protein in the plasma of normal subjects and patients with bone disease. *J Clin Invest*. 1980;66:878-883.

Prins SH, Jørgensen HL, Jørgensen LV, Hassager C. The role of quantitative ultrasound in the assessment of bone: a review. *Clin Physiol*. 1998;18:3-17.

Risteli J, Elomaa I, Niemi S, Novamo A, Risteli L. Radioimmunoassay for the pyridinoline cross-linked carboxy-terminal telopeptide of type 1 collagen: a new serum marker of bone collagen degradation. *Clin Chem*. 1993;39:635-640.

Sakou T. Bone morphogenetic proteins: from basic studies to clinical approaches. *Bone*. 1998;22:591-603.

Schlemmer A, Hassager C, Jensen SB, Christiansen C. Marked diurnal variation in urinary excretion of pyridinium cross-links in premenopausal women. *J Clin Endocrinol Metab*. 1992;74:476-480.

Seeman, E. Growth in bone mass and size—are racial and gender differences in bone mineral density more apparent than real? *J Clin Endocrinol Metab.* 1998;83:1414-1419. Editorial.

Seyedin SM, Kung VT, Daniloff YN, et al. Immunoassay for urinary pyridinoline: the new marker of bone resorption. *J Bone Miner Res.* 1993;8:635-641.

Seibel MJ, Gartenberg F, Silverberg SJ, Ratcliffe A, Robins SP, Bilezikian JP. Urinary hydroxypyridinium cross-links of collagen in primary hyperparathyroidism. *J Clin Endocrinol Metab.* 1992;74:481-486.

Simonet WS, Lacey DL, Dunstan CR, et al. Osteoprotegerin: a novel secreted protein involved in the regulation of bone density. *Cell.* 1997;89:309-319.

St. Arnaud R, Florieux FH. 24, 25-dihydroxyvitamin D-active metabolite or inactive catabolite? *Endocrinology.* 1998;139:3371-3374. Editorial.

Steinbach G, Grunert A. Procalcitonin—a new indicator for bacterial infections. *Exp Clin Endocrinol Diab.* 1998;106:164-167.

Traianedes K, Dallas MR, Garrett IR, Mundy GR, Bonewald LF. 5-Lipoxygenase metabolites inhibit bone formation *in vitro*. *Endocrinology.* 1998;139:3178-3184.

Uebelhart D, Gineyts E, Chapuy MC, Delmas PD. Urinary excretion of pyridinium crosslinks: a new marker of bone resorption in metabolic bone disease. *Bone Miner.* 1990;8:87-96.

Weinstein RS, Jilka RL, Parfitt AM, Manolagas SC. Inhibition of osteoblastogenesis and promotion of apoptosis of osteoblasts and osteocytes by glucocorticoids—potential mechanisms of their deleterious effects on bone. *J Clin Invest.* 1998;102:274-282.

Yamamoto M, Murakami T, Nishikawa M, et al. Hypocalcemic effect of osteoclastogenesis inhibitory factor/osteoprotegerin in the thyroparathyroidectomized rat. *Endocrinology.* 1998;139:4012-4015.

2 Calcium Metabolism

Introduction

Serum and extracellular calcium concentrations in mammals are closely regulated within a narrow physiologic range that is optimal for the normal cellular functions affected by calcium in many tissues (Table 2.1). Intracellular calcium is also tightly controlled, but at concentrations several orders of magnitude lower than extracellular calcium. Intracellular calcium serves as second messenger in many signal transduction pathways. The calcium gradient across the cell membrane is maintained by calcium pumps and ionic exchangers acting through cell membranes and pores and ion porters. Calcium enters cells through voltage- or receptor-operated calcium channels in respective response to membrane depolarization or receptor activation. Stores of cellular calcium in microsomal and mitochondrial pools can also be released to increase intracellular cytosolic calcium.

Extraskeletal calcium accounts for only 1% of the total body calcium that is primarily sequestered in bone. Approximately 10,000 mg/day is filtered at the glomerulus and most is reabsorbed by the renal tubules, with only a few hundred milligrams appearing in urine each day. About 500 mg is absorbed from the diet, and the unabsorbed component appears in the stool. The skeleton turns over about 250 mg/day of calcium, but there is wide variation. And in disease states, the turnover can be increased or decreased (Figure 1.1).

Approximately 50% of the total calcium in serum is ionized, with the rest bound primarily to albumin or complexed with counter-ions. The ionized calcium concentration averages 1.25 ± 0.07 mmol/L and the total serum calcium concentrations range from 8.5 to 10.5 mg/dL. Since ionized calcium has the primary regulatory role (Table 2.1), it is in turn the regulated component that maintains homeostasis. This regulation takes place through the complex interactions at their target organs of the major calcium regulating hormones, parathyroid hormone (PTH), calcitonin (CT), and vitamin D and its metabolites.

TABLE 2.1 — METABOLIC FUNCTIONS OF CALCIUM
• Neural transmission • Enzymatic reactions • Membrane function • Action potentials • Muscle contraction, motility • Secretion • Blood coagulation • Membrane potential • Exocytosis • Acts on calcium sensor

Parathyroid Hormone

Parathyroid hormone is an 84-amino-acid peptide secreted by two pairs of parathyroid glands located adjacent to the back of the thyroid gland in the neck. There can be ectopic parathyroid glands along their developmental route between the thyroid gland and mediastinum. Its relatively simple gene, located on chromosomal eleven, encodes a precursor preproPTH with a 29-amino-acid extension peptide at the amino terminus that contains a 23-amino-acid signal sequence (the "pre" sequence) and a six-residue prohormone sequence. The signal sequence serves to guide PTH to the endoplasmic reticulum and through the membrane into the lumen, where a signal peptidase cleaves the signal sequence from preproPTH to leave proPTH. ProPTH then travels to the Golgi apparatus, where the "pro" sequence is cleaved from PTH by peptidases like furin. Mature PTH is then packaged into dense secretory granules for regulated secretion.

■ Secretory Regulation of Parathyroid Hormone and the Calcium Sensor

Serum calcium inversely affects PTH secretion, with the steep portion of the sigmoidal response curve corresponding to the normal range of both. There is hysteresis in this relationship during recovery from hypo- and hypercalcemia. The parathyroid gland senses the concentration of extracellular ionized calcium through a cell-surface calcium-sensing receptor (CSR) for which calcium is an agonist.

This calcium sensor is a 120-kDa G protein-coupled receptor with seven membrane-spanning domains (heptahelical) and connecting intracellular loops that are involved in coupling the receptor to G proteins. The same sensor also regulates the responses to calcium of thyroid C cells, which secrete CT in direct relationship to extracellular calcium; the distal nephron of the kidney, where calcium excretion is regulated; the placenta, where fetal-maternal calcium fluxes occur; and the brain and gastrointestinal (GI) tract, where its function is unknown. Other sites of CSR expression are being identified, including osteoclasts and monocytes.

Genetic abnormalities of the CSR can be inactivating mutations that cause hypercalcemia in familial hypocalciuric hypercalcemia (FHH) (Chapter 3, *Hypercalcemia*) and activating mutations that cause hypocalcemia (Chapter 5, *Hypocalcemia and Hypoparathyroidism*). Mutations of the CSR are described in

Table 2.2. Mutations causing hypocalcemia are discussed in Chapter 5. CSR abnormalities, presumably acquired, have been described in primary and secondary hyperparathyroidism.

TABLE 2.2 — CALCIUM-SENSING RECEPTOR ABNORMALITIES
Inactivating
• Neonatal severe hyperparathyroidism (NSHPT)
• Familial hypocalciuric hypercalcemia (FHH)
Activating
• Familial hypoparathyroidism

Compounds have been described that act selectively on the CSR. Since they mimic or potentiate the effects of extracellular calcium at the CSR, they are termed calcimimetics. These agents, classified as phenylalkylamines and exemplified by NPSR 568, are allosteric activators of the CSR. They act on its transmembrane domain and increase sensitivity to ambient calcium, which is the receptor agonist. Receptor activation increases cytosolic calcium by redistribution of intracellular stores and efflux of extracellular calcium. This chemical parathyroidectomy not only inhibits PTH secretion but also decreases gland proliferation. Because of their corresponding renal effects, these agents can also cause hypercalciuria.

Such compounds (eg, NPSR 568) with potent and selective activity allow a pharmacological approach to regulating plasma levels of PTH that could provide a specific medical therapy for several types of hyperparathyroidism, including primary (Chapter 4, *Primary Hyperparathyroidism*), secondary (Chapter 9, *Renal Osteodystrophy*), and carcinomatous (Chapter 3, *Hypercalcemia*).

An increase in ionized calcium inhibits PTH secretion by increasing intracellular calcium through the release of calcium from intracellular stores and the influx of extracellular calcium through cell membranes and channels. This mechanism differs from most cells, where secretion of their product is stimulated by increased calcium. Intracellular magnesium may serve this secretory function in the parathyroids in that hypermagnesemia can inhibit PTH secretion and hypomagnesemia can stimulate PTH secretion. However, prolonged depletion of magnesium will inhibit PTH biosynthesis and secretion, as it will the function of many cells. Hypomagnesemia attenuates the biological effect of PTH by interfering with its signal transduction. Serum calcium also inversely regulates transcription of the PTH gene, and increased levels of 1,25-dihydroxyvitamin D (1,25-D) inhibit PTH gene transcription.

Most studies fail to demonstrate a direct effect of serum phosphate on PTH expression, but some recent studies show that high phosphate increases PTH biosynthesis and vice versa. However, serum phosphate has an inverse effect on calcium concentration and ambient phosphate directly increases 1,25-D production. Thus, serum phosphate may directly and indirectly regulate PTH expression.

■ Metabolism and Clearance of Parathyroid Hormone

Parathyroid hormone has a circulating half-life of less than 5 minutes. The hormone is metabolized to amino-terminal and carboxyl-terminal fragments primarily in the liver, also in the kidney, and perhaps in the parathyroid gland and blood. The carboxyl-terminal fragments are cleared by glomerular filtration (GF), so they accumulate in renal failure. All of the classic biological effects of PTH are mediated by the amino terminus, PTH1-34, and likely a subpeptide of this sequence, but other fragments may have their own biologic actions.

Immunoassays of PTH can be designed to measure defined parts of the molecule. Carboxy-terminal assays may be used in selected patients, but they accumulate in renal failure and now have few indications. Assays of amino-terminal PTH are intended to measure the biologically active hormone and intact assays the native molecule. However, recent studies indicate that intact assays may also measure nonintact PTH fragments. All PTH assays demonstrate increased serum levels when there is a decrease in renal function to a creatinine clearance of 35 to 40 cc/minute. Although widely used and clinically useful, intact assays may be approximated in their diagnostic value by assays that measure the mid-molecule of PTH.

■ Biologic Effects of Parathyroid Hormone

Parathyroid hormone regulates serum calcium and phosphorus concentrations through its combined actions on bone, intestine, and kidney. The skeletal effects of PTH on bone are complex. High levels of PTH, as seen in primary and secondary hyperparathyroidism, increase osteoclastic bone resorption. Low levels, especially if delivered episodically, seem to increase osteoblastic bone formation, an effect that is being studied for osteoporosis treatment (Chapter 10, *Osteoporosis*). The skeletal effects of PTH are mediated through the osteoblast, since osteoclasts do not express the PTH receptor. However, osteoblasts communicate with osteoclasts to mediate PTH effects.

Any direct gastrointestinal (GI) effect of PTH on intestinal calcium or phosphate absorption is weak. However, PTH through its stimulating effects on the renal production of 1,25-D, discussed later, promotes the absorption of both. In the kidney, PTH increases the reabsorption of calcium, predominantly in the distal convoluted tubule, and inhibits the reabsorption of phosphate in the renal proximal tubule, causing hypercalcemia and hypophosphatemia. PTH also inhibits NA^+-H^+ antiporter activity and bicarbonate reabsorption, causing a mild hyperchloremic metabolic acidosis. The major biological effects of PTH are summarized in Table 2.3.

2

TABLE 2.3 — EFFECTS OF PARATHYROID HORMONE ON MINERAL METABOLISM

Bone   • Increases resorption   • Increases formation, especially at low doses
Kidney   • Decreases calcium excretion (clearance)   • Increases phosphorus excretion
Gastrointestinal Tract   • Increases calcium and phosphorus absorption   • Indirect effect via 1,25-D production
Blood   • Increases calcium   • Decreases phosphorus

Abbreviations: 1,25-D, 1,25-dihydroxyvitamin D.

■ The PTH/PTHrP Receptor

The PTH receptor is an 80,000-MW membrane glycoprotein of the G protein receptor superfamily with the canonical structure of a large extra cellular domain, seven consecutive membrane-spanning domains (heptahelical), and a cytoplasmic tail. PTH binds at the interstices of the extracellular and transmembrane domains and initiates signal transduction. The classic PTH receptor recognizes the amino-terminus of PTH and the homologous terminus of the parathyroid hormone-related protein (PTHrP) with indistinguishable affinity; it is therefore designated the PTH/PTHrP receptor. Both PTH and PTHrP induce the active, guanosine triphosphate (GTP)-bound state of at least two receptor-associated G proteins that respectively couple the receptor to the effector adenylyl cyclase, thereby generating cyclic adenosine monophosphate (cAMP) as a cellular second messenger and activating protein kinase A (PKA), or to the phospholipase C effector system increasing cellular IP_3 and calcium and activating protein kinase C (PKC). The various cellular effects of PTH cannot yet be clearly attributed to the specific cellular messengers, PKA and PKC, or to changes in intracellular calcium. Despite this multi-signal pathway, the cAMP-PKC pathway appears to be most important for PTH/PTHrP signal transduction. In addition to this shared receptor, there is accumulating evidence for the existence of receptors that are respectively specific for PTH and PTHrP and for some of their subpeptides.

Two genetic diseases of the PTH/PTHrP receptor have been described, Jansen's chondrodysplasia, characterized by an activating mutation and Blomstrand's chondrodysplasia, characterized by absence of a functional receptor. The former has delayed endochondral bone formation and hypercalcemia; the latter, advanced endochondral bone maturation and hypocalcemia.

Parathyroid Hormone-Related Protein

Originally discovered as a product of breast cancer cells that produced hypercalcemia, PTHrP is a product of many normal and malignant tissues. PTHrP is secreted by many types of malignant tumors and produces hypercalcemia by activating the PTH/PTHrP receptor, notably by breast and lung cancer. PTHrP is produced in many fetal and adult tissues (Table 2.4). PTHrP is required for normal development as a regulator of the proliferation and mineralization of cartilage cells and as a regulator of placental calcium transport. The former effect may be recapitulated in osteoarthritis. In postnatal life, PTHrP

TABLE 2.4 — PARATHYROID HORMONE-RELATED PROTEIN EXPRESSION AND ACTION	
Expression	**Regulatory Action**
Cartilage	Chondrogenesis
Placenta and ammion	Calcium transport
Parathyroid	Secretion
Heart	Contraction/relaxation
Smooth muscle	Relaxation/contraction
Pancreas	Insulin secretion
Breast	Morphogenesis/lactation
Skin	Hair follicle development

appears to regulate the epithelial-mesenchymal interactions that are critical for development of the mammary gland, skin, and hair follicle.

The amino terminus of PTHrP reacts with the PTH/PTHrP receptor and produces most of the biological effects of native PTH, including hypercalcemia. (A second family of PTH-specific receptors has recently been identified.) The PTHrP gene expresses three forms of polypeptide through alternate messenger ribonucleic acid (mRNA) splicing:

- PTHrP 1-141
- A truncated 139 residue form
- A human-specific 173 residue form.

Whereas PTHrP 1-139 is structurally quite similar to PTHrP 1-141, PTHrP 1-173 completely diverges from both at its own carboxy terminus. The PTHrP gene sequence spans more than 15 kilobases and is composed of three primary regions:

- A complex promoter region
- A coding region
- A multiple 3'-noncoding region.

The promoter region contains three different promoter elements, designated P1, P2, and P3. P1 and P3 are "TATA box"-like, and the P2 element is a guanine cytosine (GC)-rich region.

In addition to mRNA splicing, processing of PTHrP into peptides is an important regulatory mechanism. Distinct biological properties have been attributed to the different PTHrP peptides, and specific receptors and effects have been identified. For example, PTHrP 1-34 and 1-36 mediate the growth-regulating and hypercalcemic effects of the molecule, PTHrP 35-94 promotes placental calcium transfer, and peptides included in PTHrP 109-141 inhibit osteoclast function. PTHrP also contains a nuclear localizing sequence at residues 87-107. Each of the PTHrP polypeptides can be derived through processing of native PTHrP, which contains several endoproteolytic processing sites.

Although multiple, the functions of PTHrP in malignant and normal tissues seem to be growth- and proliferation-related. There is, in fact, evidence that PTHrP interacts in its growth-regulating properties with other potent oncogenes. Furthermore, there is considerable evidence that PTHrP has direct or indirect growth-regulating activities in a variety of cells and tissues. Many factors regulate PTHrP expression, including:

- Cytokines
- Growth factors
- Hormones.

In most physiologic circumstances, PTHrP carries out local rather than systemic actions. When produced in excess by malignancy, PTHrP has systemic effects, especially hypercalcemia. Because of its protean and developmental effects, PTHrP can be considered an oncofetal protein. The regulation of PTHrP expression is thus complex (Table 2.5).

TABLE 2.5 — REGULATION OF PARATHYROID HORMONE-RELATED PROTEIN EXPRESSION
Stimulate • Angiotensin II • Cyclic adenosine monophosphate (cAMP) • Cycloheximide • Epidermal growth factor (EGF) • Transforming growth factor-beta (TGF-β) • 1,25-Dihydroxyvitamin D (1,25-D) • Estradiol • Prolactin • Sodium butyrate • *tax* • Stretch • Calcium • Interleukin 2 (IL-2) • Retinoic acid
Inhibit • Glucocorticoids • 1,25-D (bimodal) • Testosterone

Calcitonin

Calcitonin is a 32-amino acid peptide whose main effect is to inhibit osteoclast-mediated bone resorption. CT is secreted by parafollicular C cells of the thyroid and other neuroendocrine cells. In contrast to PTH, hypercalcemia increases secretion of CT while hypocalcemia inhibits secretion. CT secretion is controlled by serum calcium through the same CSR that regulates PTH secretion, but in an inverse manner and at higher concentrations of calcium or calcimimetics. The major effect of CT is to inhibit osteoclastic bone resorption. Within minutes of exposure to CT, the osteoclast shrinks in size and retracts from the bone surface. CT inhibits the renal reabsorption of phosphate, thus promoting renal phosphate excretion. CT also induces a mild natriuresis and calciureses, the latter contributing to its hypocalcemic effect.

The CT gene through alternative exon splicing and polypeptide processing ultimately encodes two peptide products, CT in thyroid C-cells which is processed from a 141-amino acid precursor, and a 37-amino peptide called gene-related peptide (CGRP) in neural tissues which is processed from a 128-amino acid precursor (Figure 2.1). CGRP is weakly recognized by the CT receptor and thereby has a CT-like effect on osteoclasts and osteoblasts. CGRP also acts through its own receptor to produce vasodilation and to act as a neurotransmitter. Calcitonin and CGRP have structural and functional homologies to amylin and adrenomedulin. This calcitonin "family" may mediate effects through related receptors whose ligand specificity is modulated by receptor-activity-modifying proteins (RAMPs) and receptor component proteins (RCPs). The main biological effects of CT are summarized in Table 2.6.

The CT receptor, like the PTH and calcium-sensing receptor, is a heptahelical G protein-coupled receptor coupled to the PKA, PKC, and Ca^{++} signal transduction pathways. Several isoforms of the receptor have been identified at different organ sites. They may play a tissue-specific role in RAMP- and RCP-mediated effects.

In addition to its role in calcium and skeletal metabolism, CT is important as a tumor marker in medullary thyroid carcinoma (see Chapter 6, *Multiple Endocrine Neoplasias*) and as an inhibitor of osteoclastic bone resorption used to treat osteoporosis (see Chapter 10, *Osteoporosis*), Paget's disease (see Chapter 8, *Paget's Disease of Bone* [*Osteitis Deformans*]), and hypercalcemia (see Chapter 3, *Hypercalcemia*).

■ Metabolism and Activation

Vitamin D is a secosterol hormone that is present in humans in an endogenous (vitamin D_3) and exogenous (vitamin D_2) form. The endogenous form of vitamin D, cholecalciferol (vitamin D_3), is synthesized in the skin from the cholesterol metabolite 7-dehydrocholesterol under the influence of ultraviolet radiation. The exogenous form of vitamin D_2 (ergocalciferol) is produced by ultraviolet irradiation of the plant sterol ergosterol and is available through the diet. Both forms of vitamin D require further metabolism to be activated, and their respective metabolism is indistinguishable. Since they are fat-soluble, they are transported in blood by specific vitamin D-binding proteins.

In the liver, vitamin D is converted by an hydroxylase to 25-hydroxyvitamin D (25-D), the principal fat storage form of vitamin D. Thus, the serum level of 25-D is the best measure of overall vitamin D status. The normal range is 10 to 50 ng/mL (25 to 125 nmol/L), but a concentration < 15 ng/mL suggests vitamin D deficiency. In the proximal tubule of the kidney, 25-D is 1α-hydroxylated to produce 1,25-D, the most active form of the hormone. The more important animal form is referred to as 1,25-dihydroxycholecalciferol. This hydroxylation step is regulated by several factors, the most important of which are PTH, phosphorus, serum calcium, and 1,25-D itself (Table 2.7). The 1α-hydroxylase that mediates this conversion in the kidney is also produced in the placenta and in keratinocytes. In certain disease states, macrophages and lymphocytes overexpress 1a-hydroxylase and produce hypercalcemia (see Chapter 3, *Hypercalcemia*).

The normal serum concentration of 1,25-D is 20 to 60 pg/mL (50 to 150 pmol/L). The kidney can also

FIGURE 2.1 — EXPRESSION OF THE CALCITONIN GENE

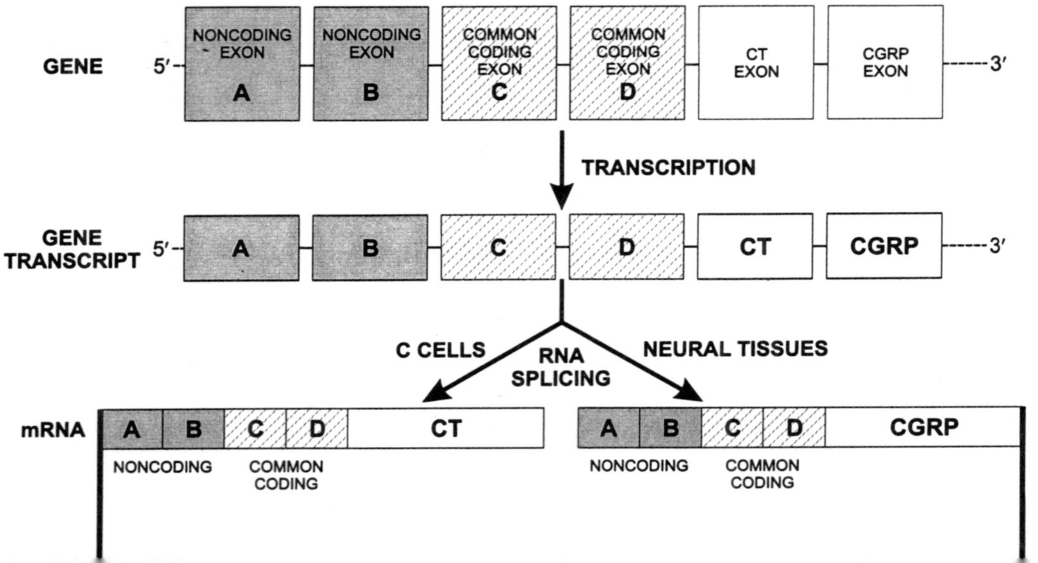

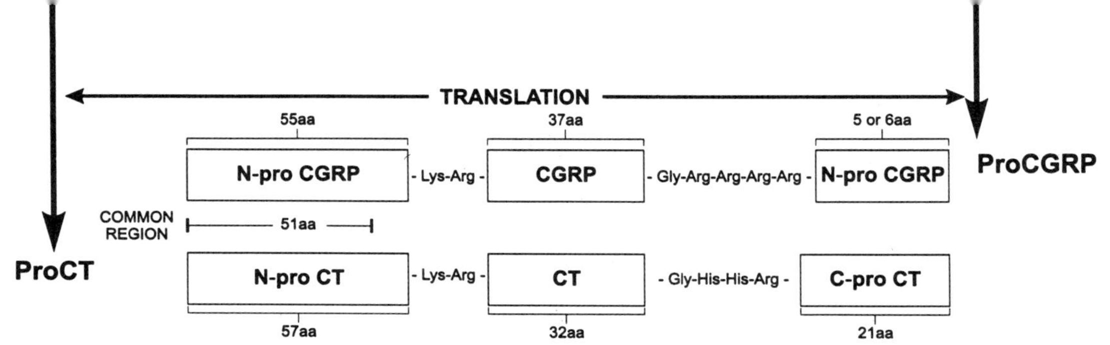

Abbreviations: CT, calcitonin; CGRP, calcitonin gene-related peptide; N-pro, amino terminal propeptide; C-pro, carboxy terminal propeptide; Lys, lysine; Arg, arginine; Gly, glycine; His, histidine.

Summary of CT gene expression and processing of its polypeptides. The two main pathways – C-cell and neural tissues – result in the respective production of CT and CGRP. The biological significance of the other processed peptides is not known.

TABLE 2.6 — EFFECTS OF CALCITONIN ON MINERAL METABOLISM
Bone • Inhibits resorption • ? Promotes formation
Kidney • Increases calcium excretion • Increases phosphorus excretion
Gastrointestinal Tract • ? Inhibitory effect on calcium/phosphorus absorption
Blood • Decreases calcium • Decreases phosphorus

TABLE 2.7 — REGULATION OF RENAL 1,25-D PRODUCTION
Stimulate • Parathyroid hormone (PTH) • Low calcium • Low phosphate • Estrogen/androgens • Prolactin • Placental lactogen • Growth hormone • Calcitonin (CT) • Insulin
Inhibit • 1,25-Dihydroxyvitamin D (1,25-D) • High calcium • High phosphate

convert 25-hydroxyvitamin D to 24,25-dihydroxyvitamin D. Although this metabolite circulates at 100-fold higher than the concentration of 1,25-D, its biologic role is unclear. Some studies suggest that it is a degradation product with no important biological effects; others suggest that it is important in chondrogenesis and bone formation, especially intramembranous. Vitamin D and its metabolites are inactivated in the liver by conjugation to glucuronides or sulfates and oxidation of their side chains.

2

The requirements for vitamin D are usually satisfied by endogenous synthesis, and milk supplementation with vitamin D makes dietary vitamin D deficiency uncommon in the United States. However, in the winter months, especially in northern climates, the institutionalized elderly may have borderline serum levels of 25-D and may be clinically vitamin D deficient, with 15 ng/dL considered borderline. The metabolism of vitamin D is summarized in Figure 2.2.

■ Biological Effects

Mechanism of Action

Vitamin D mediates its biological effects through its own member of the nuclear hormone receptor superfamily, the vitamin D receptor (VDR). The receptor binds many vitamin D metabolites with affinities that generally mirror their biological effects, and 1,25-D thus has the highest affinity. The VDR regulates gene transcription by homodimerization and by heterodimerization to a retinoic acid x receptor (RXR) specific for 9-cis-retinoic acid. The complex binds to target DNA sequences and regulates the transcription of several genes important in mediating its effects on calcium and skeletal metabolism (Table 2.8) and vitamin D's diverse biological effects (Table 2.9). Vitamin D metabolites, as well as other steroid hormones, may also act through a membrane receptor (Figure 2.2).

Intestinal Calcium Absorption

Vitamin D increases intestinal calcium absorption, primarily in the jejunum and ileum, by increasing calcium uptake through the brush border membrane of

FIGURE 2.2 — THE METABOLIC ACTIVATION OF VITAMIN D

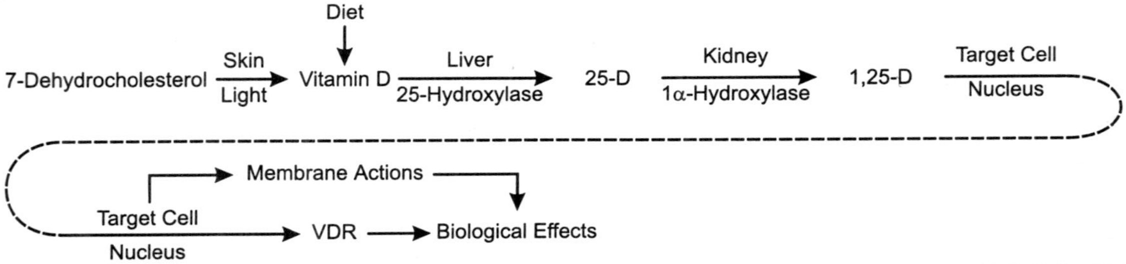

Abbreviations: 25-D, 25-hydroxyvitamin D; 1,25-D, 1,25-dihydroxyvitamin D; VDR, vitamin D receptor.

Vitamin D from the diet or the conversion from precursors in skin through ultraviolet radiation (light) provides the substrate of the indicated steps in metabolic activation. The pathways apply to both the endogenous animal form of vitamin D (vitamin D_3, cholecalciferol) and the exogenous plant form of vitamin D (vitamin D_2, ergocalciferol), both of which are present in humans at a ratio of approximately 2:1. In the kidney, 25-D is also converted to 24-hydroxylated metabolites which may have unique effects on chondrogenesis and intramembranous ossification. The many effects (Table 2.8) of vitamin D metabolites are mediated through nuclear receptors or effects on target-cell membranes.

TABLE 2.8 — EFFECTS OF 1,25-D ON MINERAL METABOLISM
Bone • Promotes mineralization of osteoid • Increases resorption at high doses
Kidney • Decreases calcium excretion • Decreases phosphorus excretion
Gastrointestinal Tract • Increases calcium absorption • Increases phosphorus absorption
Blood • Increases calcium • Increases phosphorus
Abbreviations: 1,25-D, 1,25-dihydroxyvitamin D.

2

TABLE 2.9 — DIVERSE EFFECTS OF VITAMIN D METABOLITES

Target Cell/Tissue	Action
Immune	Regulation of cytokines and lymphokines
Epithelium	Growth and differentiation
Reproductive	Effects on gonads, prostate and placenta
Central nervous system	Effects on selected ganglia, glial cells, and brain structures
Muscle	Effects on skeletal and cardiac muscle
Cancer [cells]	Effects on growth and differentiation of a variety of cancer cell lines

the enterocyte. For this action, vitamin D induces the calcium-binding calbindins, which participate in calcium transport across the cell, and it promotes the efflux of calcium from the basolateral side of the enterocyte into the circulation. This efflux is mediated by an adenosine triphosphate (ATP)-driven vitamin D-sensitive pump and by an Na^{+}-Ca^{2+} exchanger driven by the sodium gradient. The initial effects of vitamin D on intestinal calcium absorption occurs within minutes, so the actions of vitamin D on intestinal calcium transport may be also mediated by a membranous nongenomic receptor (Figure 2.2).

Bone

The effects of vitamin D metabolites on bone are complex. By providing sufficient ambient calcium and/or through some other unappreciated direct effect, vitamin D promotes the mineralization of osteoid. Vitamin D causes bone resorption by mature osteoclasts, but this effect is indirect, requiring cell recruitment and interaction with osteoblasts. Vitamin D regulates several bone proteins (Table 2.10). It promotes the transcription of osteocalcin and has bidirectional effects on type I collagen and alkaline phosphatase gene transcription. Vitamin D also promotes the fusion of monocytic precursors to osteoclasts.

Kidney

The VDR is robustly expressed in the kidney, and acting through it, 1,25-D stimulates renal proximal phosphate reabsorption and maintenance of normal calcium reabsorption. However, these effects are relatively weak.

Other Tissues

Vitamin D and its metabolites have protean effects (Table 2.9). Although vitamin D has many *in*

TABLE 2.10 — PROTEINS INDUCED BY 1,25-DIHYDROXYVITAMIN D
• Calbindin D9K • Calbindin D28K • Alkaline phosphatase • Carbonic anhydrase • Osteocalcin • Osteopontin • Matrix Gla protein • Collagen • Fibronectin • 25-Hydroxyvitamin D_3-1α-hydroxylase • 25-Hydroxyvitamin D_3-24-hydroxylase • Parathyroid hormone • Interleukin 2

vitro effects on the immune system, no major immune defect is apparent in individuals who are deficient or who lack vitamin D or its receptor. Vitamin D also inhibits proliferation and stimulates maturation of epidermal keratinocytes, which robustly express the VDR. This antiproliferative effect is being used for the treatment of psoriasis, a hyperproliferative skin disorder. Since many persons who lack vitamin D receptors have lifelong alopecia totalis, vitamin D may play a role in the maturation of the hair follicle.

Other Hormones

In addition to the primary calcemic hormones, other hormones play an important role in calcium and skeletal metabolism. Gonadal steroids maintain skeletal mass (see Chapter 10, *Osteoporosis*). Glucocorticoids are deleterious to all skeletal functions (Chapter 10). Insulin, growth hormone, and thyroid hormones promote skeletal growth and maturation. Excess production of the latter can cause hypercalcemia (see Chapter 3, *Hypercalcemia*).

Summary

Through their actions and interactions on bone, kidney and the gastrointestinal (GI) tract, the three calcitropic hormones, parathyroid hormone (PTH), calcitonin (CT), and vitamin D metabolites, especially the 1,25-dihydroxyvitamin D (1,25-D) metabolite, act to maintain serum (and extracellular fluid) calcium within a normal range, a range that optimally subserves the many calcium-requiring physiological functions. A perturbation in serum calcium, which plays an important role in regulating the concentrations of the calcitropic hormones, will cause a homeostatically appropriate and often reciprocal change in the secretion of PTH by the parathyroid glands, CT by the thyroidal C cells, and in the production of 1,25-D by the kidneys. These responses are designed to return the serum calcium, and, to a lesser extent, the serum phosphorus and magnesium to normal.

Thus, the patient with hypocalcemia (nonparathyroid) will have an increased serum PTH and 1,25-D and a decreased serum CT. This will result in increased GI absorption of calcium, increased bone resorption, and decreased renal calcium excretion – all acting to increase the serum calcium toward normal. The patient with hypercalcemia (nonparathyroid) will have a decreased serum PTH and 1,25-D and an increased serum CT. This will result in decreased GI absorption of calcium, decreased bone resorption, and increased renal calcium excretion – all acting to decrease the serum calcium toward normal. As is always the case with compensatory mechanisms, the normal state, here of serum calcium, will not be attained until the primary abnormality is corrected.

In diseases that involve one of the three calcitropic hormones, the serum concentrations of the other two will change to either amplify the effect of the primary abnormality or to defend against the calcium perturbation (see Chapter 4, *Primary Hyperparathyroidism*; Chapter 6, *Multiple Endocrine Neoplasias*; and Chapter 7, *Osteomalacia and Rickets*). Although these compensatory mechanisms act to restore serum calcium to normal, the homeostasis will not be complete until the primary abnormality has been corrected. In addition to these three calcitropic hormones, other hormones, cytokines, and growth factors play an important role in calcium metabolism. Among the other important hormones are insulin, growth hormone, and the gonadal and adrenal steroids and thyroid hormone.

2

SUGGESTED READING

Bell NH. Renal and nonrenal 25-hydroxyvitamin D-1α-hydroxylases and their clinical significance. *J Bone Miner Res*. 1998; 13:350-353.

Brown EM, Hebert SC. Calcium-receptor-regulated parathyroid and renal function. *Bone*. 1997;20:303-309.

Brown EM, Pollak M, Hebert SC. The extracellular calcium-sensing receptor: its role in health and disease. *Annu Rev Med*. 1998;49:15-29.

Chapuy MC, Arlot ME, Duboeuf F, et al. Vitamin D3 and calcium to prevent hip fractures in elderly women. *N Engl J Med*. 1992;327:1637-1642.

Deftos LJ. There's something fishy and perhaps fowl about the mammalian calcitonin receptor and its ligand. *Endocrinology*. 1997;138:519-520. Editorial.

Deftos LJ, Gazdar AF, Ikeda K, Broadus AE. The parathyroid hormone-related protein associated with malignancy is secreted by neuroendocrine tumors. *Mol Endocrinol*. 1989;3:503-508.

Deftos LJ, Weisman MH, Williams GW, et al. Influence of age and sex on plasma calcitonin in human beings. *N Engl J Med*. 1980;302:1351-1353.

Goodman WG, Veldhuis JD, Belin TR, Van Herle AJ, Juppner H, Salusky IB. Calcium-sensing by parathyroid glands in secondary hyperparathyroidism. *J Clin Endocrinol Metab.* 1998;83:2765-2772.

Haussler MR, Whitfield GK, Haussler CA, et al. The nuclear vitamin D receptor: biological and molecular regulatory properties revealed. *J Bone Miner Res.* 1998;13:325-349.

Iirl T, Farfel Z, Bourne HR. G-protein diseases furnish a model for the turn-on switch. *Nature.* 1998;394:35-38.

Jobert AS, Zhang P, Couvineau A, et al. Absence of functional receptors for parathyroid hormone and parathyroid hormone-related peptide in blomstrand chondrodysplasia. *J Clin Invest.* 1998;102: 34-40.

Knecht TP, Behling CA, Burton DW, Glass CK, Deftos LJ. The humoral hypercalcemia of benignancy. A newly appreciated syndrome. *Am J Clin Pathol.* 1996;105:487-492.

Kovacs CS, Ho-Pao CL, Hunzelman JL, et al. Regulation of murine fetal-placental calcium metabolism by the calcium-sensing receptor. *J Clin Invest.* 1998;101:2812-2820.

Manolagas SC, Deftos LJ. The vitamin D endocrine system and the hematolymphopoietic tissue. *Ann Intern Med.* 1984;100:144-146.

McLatchie LM, Fraser NJ, Main MJ, et al. RAMPs regulate the transport and ligand specificity of the calcitonin-receptor-like receptor. *Nature.* 1998;393:333-339.

Moallem E, Kilav R, Silver J, Naveh-Many T. RNA-protein binding and post-transcriptional regulation of parathyroid hormone gene expression by calcium and phosphate. *J Biol Chem.* 1998;273:5253-5259.

Morinaga T, Nakagawa N, Yasuda H, Tsuda E, Higashio K. Cloning and characterization of the gene encoding human osteoprotegerin/osteoclastogenesis-inhibitory factor. *Eur J Biochem.* 1998;254: 685-691.

Muff R, Born W, Fischer JA. Calcitonin, calcitonin gene-related peptide, adrenomedullin and amylin: homologous peptides, separate receptors and overlapping biological actions. *Eur J Endocrinol.* 1995;133:17-20.

Nemere I, Farach-Carson MC. Membrane receptors for steroid hormones: a case for specific cell surface binding sites for vitamin D metabolites and estrogens. *Biochem Biophys Res Commun.* 1998;248:443-449.

Nguyen MTA, Karaplis AC. The nucleus: a target site for parathyroid hormone-related peptide (PTHrP) action. *J Cell Biochem.* 1998;70:193-199.

Pearce SHS. Endocrinology: calcium homeostasis and disorders of the calcium-sensing receptor. *J R Coll Physicians Lond.* 1998; 32:10-13.

Philbrick WM. Parathyroid hormone-related protein is a developmental regulatory molecule. *Eur J Oral Sci.* 1998;106(suppl 1):32-37.

Riggs BL, O'Fallon WM, Muhs J, O'Connor MK, Kumar R, Melton LJ III. Long-term effects of calcium supplementation on serum parathyroid hormone level, bone turnover, and bone loss in elderly women. *J Bone Miner Res.* 1998;13:168-174.

Schwarz P, Madsen JC, Rasmussen AQ, Transbøl I, Brown EM. Evidence for a role of intracellular stored parathyroid hormone in producing hysteresis of the PTH-calcium relationship in normal humans. *Clin Endocrinol.* 1998;48:725-732.

Schwindinger WF, Fredericks J, Watkins L, et al. Coupling of the PTH/PTHrP receptor to multiple G-proteins. *Endocrine.* 1998;8: 201-209.

St-Arnaud R, Glorieux FH. 24, 25-Dihydroxyvitamin D – active metabolite or inactive catabolite? *Endocrinology.* 1998;139:3371-3374. Editorial.

Suzuki H. Calcitonin levels in normal individuals with new highly sensitive chemiluminescent enzyme immunoassay. *J Clin Lab Anal.* 1998;12:218-222.

Taylor AN, Inpanbutr N, Deftos LJ. Localization of calbindin-D28K in calcitonin containing cells of chick ultimobranchial glands. *Anat Rec.* 1987;219:86-90.

Zhu W, Malloy PJ, Delvin E, Chabot G, Feldman D. Hereditary 1,25-dihydroxyvitamin D-resistant rickets due to an opal mutation causing premature termination of the vitamin D receptor. *J Bone Miner Res.* 1998;13:259-264.

PART II

DISEASES OF CALCIUM METABOLISM

3

Hypercalcemia

Introduction

3

Hypercalcemia is one of the most common metabolic disorders. The two most common causes of hypercalcemia are primary hyperparathyroidism (PH) and malignancy, and they account for the vast majority of cases (Table 3.1). Hypercalcemia can occur through several mechanisms, with increased osteoclastic bone resorption being the most common. PH is usually seen in the clinic and malignancy in the hospital. Although there are exceptions, serum calciums of 11 to 12 mg/dL are often asymptomatic while calciums greater than 12 or rapidly rising are likely to cause symptoms. In general, the signs and symptoms of PH are usually subtle, while those of malignancy dominate the clinical picture. However, where health care is poor, the classical signs and symptoms of primary hyperparathyroidism will be manifest. The symptoms of hypercalcemia are related to the effects of extracellular calcium on the function of many tissues, including:

- Brain
- Peripheral nerves
- Visceral smooth muscle
- Cardiac muscle
- Kidney.

The major clinical manifestations are:

- Lethargy
- Impaired mentation
- Confusion
- Anorexia

TABLE 3.1 — DIFFERENTIAL DIAGNOSIS OF HYPERCALCEMIA
Artifactual • Hyperproteinemia (dehydration) • Venous stasis (during blood collection) • Hyperalbuminemia (eg, hyperalimentation) • Hypergammaglobulinemia (eg, myeloma, Waldenström's syndrome) • Hyperlipidemia • Cork-stoppered tubes
Malignancy • Skeletal metastases (local osteolysis) • "Humors" (Table 3.2)
Drugs • Vitamin D (including topical analogs) • Vitamin A • Calcium • Lithium • Manganese • Thiazides • Theophylline • Tamoxifen (anti-estrogens and estrogens) • Growth hormone therapy • Parenteral nutrition • Milk alkali syndrome (calcium and vitamin D) • Dialysis (calcium)
Endocrine • Primary hyperparathyroidism (PH) • Multiple endocrine neoplasias (MEN) • Zollinger-Ellison (ZE) syndrome • Familial hyperparathyroidism (FH) • Familial hypocalciuric hypercalcemia (FHH) • Neonatal severe hyperparathyroidism (NSHPT) • Idiopathic hypercalcemia of infancy • Hyperthyroidism • Hypoandrenalism • Pheochromocytoma • Vipoma

Renal Disease • Diuretic phase of acute renal failure • "Tertiary" hyperparathyroidism
"Granulomatous" Disease • Sarcoidosis • Coccidioidomycosis • Paracoccidioidomycosis • Tuberculosis • Histoplasmosis • Berylliosis • Brucellosis • Leprosy • Nocardia • Candidiasis • Silicone injections or implants • Cat-scratch disease • Pulmonary eosinophilic granuloma • Wegener's granulomatosis • Pneumocystis • Crohn's disease
Immobilization, especially with: • Paget's disease • Adolescence • Thyrotoxicosis
Pediatric Disorders • Infantile hypercalcemia (William's syndrome) • Hypophosphatasia • Jansen's metaphyseal chondrodysplasia
Other • Systemic lupus erythematosis • Acquired immunodeficiency syndrome (AIDS) • Rheumatoid arthritis • Advanced chronic liver disease

- Nausea
- Vomiting
- Constipation
- Polyuria
- Polydipsia.

The symptoms correlate in a general way with the severity, speed of development, and duration of the hypercalcemia. They respond to effective treatment if due to the hypercalcemia in a given case, but symptomatic improvement lags behind serum calcium improvement. A number of complications may result from the hypercalcemic state. These include:

- Dehydration
- Nephrolithiasis
- Nephrocalcinosis
- Renal failure
- Hypertension
- Cardiac
- Arrhythmias
- Peptic ulceration
- Pancreatitis
- Coma.

A common consequence of hypercalcemia is dehydration. In addition, other effects of hypercalcemia, such as vomiting or gastrointestinal (GI) bleeding, may contribute to dehydration. Consequent decreased glomerular filtration and, in turn, decreased renal calcium clearance exacerbate hypercalcemia, thus creating a vicious hypercalcemic cycle.

Primary Hyperparathyroidism

Primary hyperparathyroidism is one of the most common causes of hypercalcemia. It is discussed in Chapter 4, *Primary Hyperparathyroidism*.

Malignancy and Hypercalcemia

The hypercalcemia of malignancy is usually due to increased bone resorption that is caused by:

- Skeletal metastases
- Production by the tumor of a "humor" that stimulates osteoclastic bone resorption.

It is likely that the first mechanism also involves the second, since most tumor cells do not have the capacity to directly resorb bone. Many cell types and their products participate in pathogenesis (Figure 3.1), and many tumor products that stimulate osteoclastic bone resorption have been implicated in the pathogenesis of the hypercalcemia of malignancy (Table 3.2). The most common seems to be the parathyroid hormone-related protein (PTHrP) (see Chapter 2, *Calcium Metabolism*), especially in solid tumors where PTHrP expression can be implicated in up to 80% of patients. Originally discovered as a product of malignant cells that produce hypercalcemia, PTHrP has been demonstrated to be a product of many normal and malignant tissues (Table 2.4).

Parathyroid hormone-related protein overexpression occurs commonly in cancers of the lung and breast and in adult T-cell leukemia and lymphoma, but it has been reported in most cancers. The measurement in serum of PTHrP by immunoassay can be used to document the presence of a PTHrP-producing tumor. However, currently available assays are relatively insensitive as to miss the diagnosis in many patients. The measurement of a low PTH (PTHrP or parathyroid hormone [PTH] assays do not cross-react) in a hypercalcemic patient indicates a non-PTH cause of hypercalcemia and can be presumptive of a PTHrP etiology in the patient with cancer since it is the most common "humoral" mediator of the hyper-

FIGURE 3.1 — PATHOGENESIS OF THE HYPERCALCEMIA OF MALIGNANCY

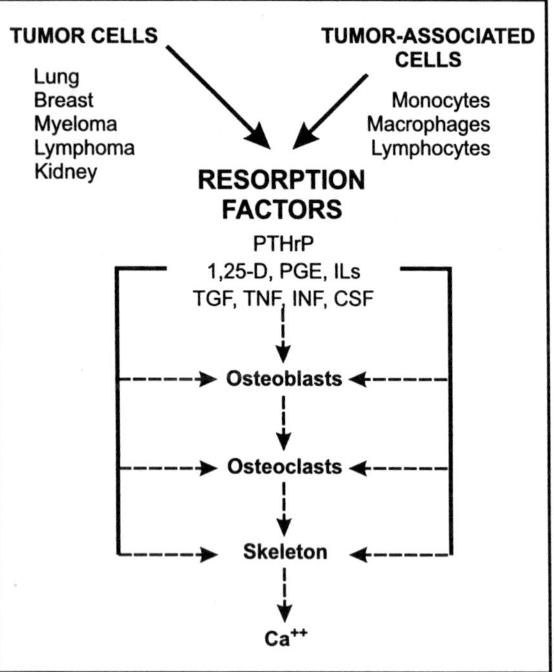

Abbreviations: PTHrP, parathyroid hormone-related protein; 1,25-D, 1,25-dihydroxyvitamin D; PGE, prostaglandin E; IL, interleukin; TGF, transforming growth factor; TNF, tumor necrosis factor; INF, interferon; CSF, colony-stimulating factor; Ca^{++}, calcium.

calcemia of malignancy. The amino terminus of PTHrP binds with the PTH receptor and produces most of the biological effects of native PTH, including hypercalcemia. (This receptor is thus referred to as the PTH/PTHrP receptor.)

Other cell products are also likely to play a causal role in the hypercalcemia, including tumor necrosis factors (TNFs), interleukins (IL-6 and IL-11 in addi-

TABLE 3.2 — TUMOR AND IMMUNE CELL PRODUCTS IN THE PATHOGENESIS OF "HUMORAL" HYPERCALCEMIA OF MALIGNANCY

• Parathryoid hormone-related protein (PTHrP)
• Parathyroid hormone (PTH) – rare
• 1-25-Dihydroxyvitamin D (1,25-D)
• Transforming growth factor-alpha and -beta (TGF-α and -β)
• Tumor necrosis factor-alpha and -beta (TNF-α and -β)
• Interleukin 1 (IL-1)
• IL-6
• IL-11
• Granulocyte-macrophage colony-stimulating factor (GM-CSF)
• Prostaglandin E_2 (PGE_2)
• Interferons (INFs)

tion to IL-1), the interferons (INFs), and colony-stimulating factors (CSFs). These can be produced by the tumor cells or by the tumor-associated immune cells. 1,25-D is abnormally produced by certain hematopoietic cancers. Transforming growth factor-beta (TGF-β) can also participate in pathogenesis by stimulating PTHrP production from tumors or immune cells as it is released from its skeletal reservoir upon osteoclastic resorption. Excess prostaglandin E_2 production by cancers is no longer considered common. Ectopic PTH production is now recognized as rare, with approximately one half dozen well documented cases.

Other Causes of Hypercalcemia

■ Endocrinopathies

Primary Hyperparathyroidism

Primary hyperparathyroidism is the most common endocrine cause of hypercalcemia (see Chapter 4, *Primary Hyperparathyroidism*). While PH most commonly occurs by itself, it can also occur as a part of multiple endocrine neoplasia (MEN) (see Chapter 6, *Multiple Endocrine Neoplasias*); including the Zollinger-Ellison (ZE) syndrome, in familial hypocalciuric hypercalcemia (FHH), and as familial hyperparathyroidism (FH).

Hyperthyroidism

Mild hypercalcemia is found in a substantial minority of patients with thyrotoxicosis or excess thyroid hormone medication. The hypercalcemia is related to T_3/T_4 levels, since thyroid hormone excess stimulates osteoclastic bone resorption. Bone markers such as alkaline phosphatase can also be elevated. The hypercalcemia resolves upon successful treatment of the hyperthyroidism.

Familial Hypocalciuric Hypercalcemia

Familial hypocalciuric hypercalcemia is an autosomal dominant disease due usually to loss-of-function (inactivating) mutations of the calcium-sensing receptor (CSR). (Loss of both alleles in children of two affected parents can result in severe neonatal hypercalcemia, a life-threatening disorder.) The loss of one functional receptor allele shifts the set-point for inhibition by serum calcium of PTH secretion to a higher calcium level, producing hypercalcemia. A corresponding defect of the same receptor in the distal renal tubule likely causes the decreased renal calcium excretion and the resulting relative hypocalciuria (but not necessarily below normal), the hallmark of the disease. (Hypocalciuria can also be seen in the hypercalcemia due to thiazide and lithium administration.) Hypercalcemia is usually mild and the PTH level is normal or only slightly elevated. There can be corresponding changes in the parathyroid glands, but most reports show normal parathyroid histology. The urinary calcium level is usually less than

100 mg/24 hr (range: 30 to 80) and the calcium/creatinine clearance ratio is less than 0.10.

Diagnosis of FHH is based on family studies and/or genetic testing, with only the latter being conclusive. However, such testing is not yet clinically practical since there are more than 25 mutations, one third of patients have no CSR mutations despite mapping to the locus, and two families have been described with non-CSR mutations. With a penetrance of 100%, each first-degree relative will have a 50% chance of being hypercalcemic. Since a large number of different point mutations in the calcium sensor can produce FHH, molecular genetic diagnosis is complex at present and only available through research laboratories. Because FHH is almost always asymptomatic with a nondiagnostic serum PTH, distinguishing it from PH is important to avoid misdiagnosis and unnecessary parathyroidectomy.

Adrenal Insufficiency

Hypercalcemia can be a feature of acute adrenal crisis. Hemoconcentration may be causative, and ionized calcium is often normal despite the increase in total calcium. Acute adrenal insufficiency has also been reported to cause hypocalcemia.

■ Drugs

Thiazide Diuretics

The administration of thiazides and related diuretics can produce an increase in the serum calcium by acting on the renal tubules to decrease the renal excretion of calcium. This effect of thiazides is used to treat hypercalciuria (Chapter 11, *Hypercalciuria and Nephrolithiasis*). Thiazide administration can also exacerbate the effects of any underlying hypercalcemia disorder such as PH. The hypercalcemia is mild and usually transient, lasting for days or weeks, but occasionally it persists.

Lithium Therapy

Lithium administration can produce hypercalcemia by decreasing the renal excretion of calcium and/or by shifting the parathyroid glands' set-point for inhibition by calcium of PTH secretion to the right, causing elevated levels of PTH. Most patients with therapeutic lithium levels have a slight increase in the serum calcium level, but only a small minority actually become hypercalcemic. Causality is established by the prudent discontinuation of lithium.

Vitamins D and A

Large doses of vitamin D, usually 50,000 IU daily, can cause hypercalcemia by increasing calcium absorption and bone resorption. The serum level of 1,25-dihydroxyvitamin D (1,25-D) may be normal or sightly raised, because of the closely regulated control of its production, but serum 25-hydroxyvitamin D (25-D) levels are 5 to 10 times normal. Thus, hypercalcemia likely results from the direct effects of the elevated levels of 25-D acting on the vitamin D receptor (VDR), even though its affinity for the receptor is an order of magnitude less than that of 1,25-D. Exuberant use of topical vitamin D preparations on denuded psoriatic skin may also cause hypercalcemia.

Vitamin A intoxication is caused by large doses of vitamin A (50,000 to 200,000 IU daily). Increased osteoclastic bone resorption seems to be the primary mechanism. An increased incidence might be expected because of the growing number of vitamin A-containing medicines.

Overproduction of 1,25-D can also cause hypercalcemia. This can occur in certain granulomatous diseases, notably sarcoidosis, and certain cancers, notably lymphomas. The etiologic cells, macrophages in sarcoid and lymphoid cells in lymphoma, express the 1α-hydroxylase that affects the overproduction.

Milk-Alkali Syndrome

The ingestion of large quantities of calcium together with an absorbable alkali can produce hypercalcemia with alkalosis and inhibit renal function. The hypercalcemia is caused by increased calcium absorption through the ulcer-involved mucosa and the often-accompanying decrease in renal calcium excretion. This cause of hypercalcemia is seen when large amounts of calcium-containing absorbable antacids are used in the treatment of peptic ulcers in a patient with renal disease.

Other Drugs

Other drugs have been occasionally reported to cause hypercalcemia (Table 3.1).

■ Renal Disease

Acute Renal Failure

Hypercalcemia can be seen in the recovery phase of renal failure, especially when precipitated by rhabdomyolysis. Although the pathogenesis is uncertain, it has been postulated to be due to a persistent secondary hyperparathyroidism. The hypercalcemia is usually transient and resolves over a few weeks.

Tertiary Hyperparathyroidism

True tertiary hyperparathyroidism is extremely rare. When reported, it occurs in the setting of severe and prolonged renal failure, where the pathogenesis is presumably due to "autonomous" function of secondarily hyperplastic glands. Although apparently uncommon, there can be monoclonal expansion of the hyperplastic parathyroid cells of renal failure with consequent autonomously functioning tissue that needs surgical attention. However, hypercalcemia in renal failure is more likely due to vitamin D and/or calcium administration, and these should always be ruled out before the diagnosis of tertiary hyperparathyroidism is made and parathyroidectomy (PTX) considered. PTX makes a difficult to treat disease more difficult.

■ Granulomatous Disease

Hypercalcemia is seen in a significant fraction of patients with granulomatous disorders, especially sarcoidosis (Table 3.1). The hypercalcemia can be caused by increased conversion of 25-D to its active form, 1,25-D, by the enzyme 25-hydroxyvitamin D-1α-hydroxylase in disease-associated macrophages and immune cells. Skin testing may reveal anergy in sarcoidosis and serology may help in other granulomas such as coccidioidomycosis. Measurement of angiotensin-converting enzyme (ACE) may be of diagnostic use in some patients with sarcoidosis and other granulomas, but the test is not specific and can be increased in other chronic infections. Recently, abnormal PTHrP production has been observed in sarcoidosis. The treatment is disease specific.

■ Immobilization

The immobilization of patients with a preexisting high bone turnover state, such as adolescence, thyrotoxicosis, or Paget's disease, can cause hypercalcemia. The mechanism is likely due to the persistence of the increased bone resorption, while bone formation is attenuated by the immobilization and consequent lack of muscle tension on bone.

■ Differential Diagnosis

Although the list (Table 3.1) is long, the correct diagnosis of hypercalcemia can usually be established by history, physical exam, and a few tests. However, when the diagnosis is unclear, extensive testing may be necessary as outlined in Table 3.3.

TABLE 3.3 — SPECIAL TESTS IN THE DIFFERENTIAL DIAGNOSIS OF HYPERCALCEMIA

Artifactual
- Serum proteins
- Serum lipids
- Ionized calcium

Hyperparathyroidism
- PTH
- Serum phosphate (low in the fasting state)

Malignancy
- Parathyroid hormone-related protein (PTHrP) – many cancers
- Vitamin D metabolites:
  - Lymphomas
  - Granulomas
  - Overdose
- Serum and urine electrophoresis
  - Myeloma
  - Some lymphomas
- Calcitonin (CT) – neuroendocrine cancers
- Special tests when available (Table 3.2)

Granulomatous Disease
- Skin test (anergy) and angionten-converting enzyme (ACE) – sarcoidosis
- Serology – infections, eg, coccidioidomycosis

Endocrine (appropriate endocrine test)
- Thyrotoxicosis – T_3/T_4, thryoid-stimulating hormone (TSH)
- Multiple endocrine neoplasia (MEN) – CT, chromogranin A
- Zollinger-Ellison (ZE) syndrome – gastrin
- Familial hypocalciuric hypercalcemia (FHH) – urinary calcium
- Hypoadrenalism – cortisol, adrenocorticotropic hormone (ACTH)
- Pheochromocytoma – epinephrine, norepinephrine

Treatment of Hypercalcemia

Hypercalcemia can occur through three mechanisms and is obversely treated (Table 3.4). Increased osteoclastic bone resorption is the most common mechanism of hypercalcemia, but increased GI absorption of calcium can contribute in vitamin D-mediated hypercalcemia and the milk-alkali syndrome, and decreased renal excretion of calcium can contribute in lithium- and thiazide-related hypercalcemia and in FHH. Hypercalcemia can be effectively treated in most situations. The severity of hypercalcemia and urgency for its treatment are assessed by:

- Consideration of the patient's symptoms
- Presence of established complications
- Degree and rapidity of elevation of the serum calcium.

TABLE 3.4 — MECHANISMS IN PATHOGENESIS AND TREATMENT OF HYPERCALCEMIA
Mechanisms in the Pathogenesis of Hypercalcemia • Increased bone resorption • Increased gastrointestinal calcium absorption • Decreased renal calcium excretion
Mechanisms in the Treatment of Hypercalcemia • Inhibit bone resorption • Decrease gastrointestinal calcium absorption • Increase renal calcium excretion

The general mechanisms of the various pathogeneses and the treatment regimens to be discussed are summarized in Table 3.4. Specific treatment regimens are outlined in Table 3.5.

Volume Expansion and Calciuresis

The promotion of urinary calcium excretion by the administration of saline (volume expansion) fol-

TABLE 3.5 — TREATMENT OF HYPERCALCEMIA
Volume Expansion and Calciuresis • Hydration • Saline • Loop diuretics: – Furosemide – Ethacrynic acid
Treatment of the Primary Disease • Tumor treatment • Parathyroidectomy (PTX) • Glucocorticoids (responsive disease, eg, lymphoma)
Inhibition of Bone Resorption • Calcitonin-salmon (Miacalcin) • Bisphosphonates (pamidronate [Aredia])
Reduction of Calcium Absorption • Low calcium diet • Decrease in parathyroid hormone and vitamin D
Other • Estrogens/progestogens • Gallium nitrate (Ganite) • WR-2721 (Ethiofos) • Chloroquine • Ketoconazole • Plicamycin (Mithracin) • Prostaglandin synthetase inhibitors (indomethacin) • Mobilization • Calcium-sensor mimetics • Ethylenediamine-tetraacetic acid (EDTA) • Dialysis

lowed by diuretics (calciuresis) is the cornerstone of the emergency management of hypercalcemia. Saline can be infused carefully at 200 mL to 250 mL/hour or more, depending on:

- Severity of hypercalcemia
- Cardiovascular function
- Renal function.

Urinary clearance of several liters/day of saline is usually necessary for the adequate treatment of hypercalcemia until bone resorption can be controlled by one or more of the agents to be discussed subsequently. Although invasive, measurement of right atrial or pulmonary capillary wedge pressure should be considered to monitor fluid therapy in patients with heart failure and severe hypercalcemia. Careful recording of intakes and urinary volume and serum sodium, as well as repeated clinical evaluation, is thus essential to maintain fluid and sodium balance; monitoring pulmonary capillary wedge pressure should seriously be considered in patients with significant heart failure.

An adjunct to calciuretic therapy is a loop diuretic such as furosemide (Lasix, 100 mg/2 hr) or ethacrynic acid (Edecrin, 40 mg/2hr) to assist the renal excretion of the infused saline and with it the co-transported, filtered load of calcium. To avoid recurrent dehydration, it is important not to institute the diuretic until extracellular volume expansion has been achieved. In patients with mild intrinsic renal disease, therapy with saline and loop diuretics can be used as long as an effective response to the diuretic can be achieved. In patients with significant renal failure, this mode of therapy is not effective, and if the serum calcium must be rapidly lowered because of life-threatening cardiac or central nervous system effects, emergency dialysis may be the best means to rapidly and safely lower the serum calcium.

■ Inhibition of Bone Resorption

Calcitonin

Calcitonin (CT), usually salmon (SCT), has many clinical advantages in the treatment of hypercalcemia because it:

- Is a potent inhibitor of osteoclastic bone resorption.

- Can act rapidly to correct hypercalcemia.
- Has few serious complications.

The recommended dose is 4 to 8 IU/Kg SC every 6 or 12 hours; the maximum dose should be used. Skin testing can probably be bypassed in order to quickly start therapy unless the patient has received SCT, or there is another reason to suspect the rare allergic reaction. CT treatment does have several limitations. Although it can rapidly lower blood calcium, normal levels are seldom achieved. Furthermore, resistance (likely tachyphylaxis) to the hypocalcemic effect of CT is commonly seen after several days of treatment. Concomitant treatment with glucocorticoids has been reported to improve the initial effectiveness of CT treatment and may be worth trying in selected patients.

Bisphosphonates

This group of drugs has been an important addition to regimens for the treatment of hypercalcemia. Several newer agents of this class of drug are very effective for lowering blood calcium. These agents inhibit bone resorption by inhibiting osteoclasts and acting as stabilizing bone-crystal substitutes. Pamidronate (Aredia) is approved by the Food and Drug Administration (FDA) for this indication. It markedly lowers the serum calcium within several days and, in many patients, restores the serum calcium toward normal. The initial dose is 60 to 90 mg given by intravenous (IV) infusion. Protocols can vary, ranging from 4 to 24 hr, with the latter in the package insert. However, shorter infusion times are more convenient and appear to be as effective and safe. The nadir of serum calcium is 4 to 5 days and may persist for weeks. Oral pamidronate does not appear to be thus useful. However, other bisphosphonates such as zolendronate and ibandronate are being developed for this indication. Some of these newer agents can also be administered by IV push and thus promise more convenience, especially for outpatient use.

The FDA has recently approved the use of pamidronate in normocalcemic patients with bone metastases or myeloma, where it can have a beneficial effect on:

- Pain
- Pathological fracture
- Prevention of hypercalcemia.

Recent studies suggest that bisphosphonates can reduce the incidence of new bony and visceral metastases in breast cancer.

■ Reduction of Calcium Absorption

Diets high in calcium and vitamin D should be avoided, but low calcium diets are unpalatable, usually not effective, and may increase serum PTH. Thiazide diuretics, which decrease renal calcium excretion, should be avoided. Loop diuretics, useful in the acute treatment of hypercalcemia because of their calciuretic action, have no role in sustained medical management. Dairy products and other foods high in calcium may be moderated in hypercalcemic disorders with increased intestinal calcium absorption such as hyperparathyroidism, vitamin D intoxication, idiopathic hypercalciuria, and sarcoidosis, but strict low calcium diets are not more effective.

■ Calcimimetic Drugs

Calcimimetic drugs that block the calcium sensor on the parathyroid glands and decrease PTH secretion are under experimental evaluation. They show promise in early trials of the treatment for primary (and secondary) hyperparathyroidism and parathyroid carcinoma (Chapter 2, *Calcium Metabolism*).

Diagnosis and Treatment of Underlying Disease

Although less important in acute management, correct identification of the cause of hypercalcemia is obviously essential in the long run. The generally effective measures to inhibit bone resorption discussed above only provide days to weeks of control of the hypercalcemia, and more definitive diagnosis and treatment of the underlying etiology must be promptly undertaken.

It is important that the diagnostically crucial serum measurement of PTH be done on a pretreatment blood sample. Otherwise, an induced lowering of serum calcium may increase the PTH and cause diagnostic confusion. Corresponding changes can take place in other tests. When the primary tumor or disease fails to respond to direct treatment, control of the hypercalcemia becomes progressively more difficult, and the decision to continue antihypercalcemic treatment can be disappointing.

Glucocorticoids

Although widely recommended for hypercalcemia, glucocorticoids are not generally effective inhibitors of bone resorption and should be reserved for cases in which a specific glucocorticoid-responsive etiology for the hypercalcemia is suspected, such as:

- Vitamin D intoxication
- Sarcoidosis
- Responsive granulomatous diseases.

A second group of patients in whom glucocorticoids are indicated are those with glucocorticoid-responsive malignant disease, including some:

- Myelomas
- Lymphomas
- Lymphocytic leukemias
- Breast cancers.

In these situations, initiation of glucocorticoid therapy may be helpful prior to, or even in the absence of, other chemotherapy. However, hypercalcemia due to other solid tumors does not usually respond to this treatment.

■ Other Treatments

Oral Phosphorus

Once widely recommended, this regimen has been superseded by the other regimens discussed above and now has little clinical import. Since bone resorption is in part controlled by the extracellular calcium-phosphate product, correction of hypophosphatemia was shown to inhibit bone resorption and lower blood calcium. However, calcification of vital organs is one of the complications of hypercalcemia that is dependent on the calcium-phosphate ion product. Thus, treatment with phosphorus is not recommended unless serum phosphorus has fallen to a level where complications of hyperphosphatemia are not of concern (ie, < 2 mg/dL). If administered, it should be given only by mouth in doses of up to 1.5 g of phosphorus per day, usually a neutral phosphate solution, and renal function must be observed very closely (Table 12.7). Diarrhea usually limits oral phosphate administration. Parenteral phosphate therapy can cause severe extraskeletal calcification.

Other Agents and Regimens

Other agents are available for the treatment of hypercalcemia (Table 3.5), but their use is currently limited. Estrogens may have some use in the postmenopausal female when other treatments are not useful. Although gallium nitrate is FDA approved for hypercalcemia, it requires a prolonged treatment protocol and experience with it is limited. Chloroquine may be useful in some patients with sarcoidosis.

Plicamycin has been superseded by the bisphosphonates and indomethacin is generally not effective since prostaglandin excess is no longer considered a common etiology. Ethiofos and the calcium-sensor mimetics (calcimimetics) are still in their early stages, but the latter are promising (see Chapter 2, *Calcium Metabolism*). Mobilization is only of limited use in the immobilized patient. Ethylenediamine-tetraacetic acid (EDTA) and dialysis, with few exceptions, are no longer indicated because of the availability of the other effective and less complex regimens.

3

Other General Measures

Hypercalcemia may cause ventricular arrhythmias, especially in patients receiving digitalis glycosides. In this setting, meticulous care should be given to replacement of urinary potassium and magnesium losses and treatment of their deficiencies. Immobilization can aggravate hypercalcemia. The effect of physical immobility is most easily seen when patients with high bone turnover rates (eg, adolescents, patients with thyrotoxicosis or Paget's disease) are forced by trauma or another severe illness to observe complete bed rest. In such cases, hypercalcemia results from an abrupt decrease in the rate of calcium accretion in the face of continued bone resorption, with an increased net flux of calcium from bone to blood. Such patients should be ambulated as soon and as much as possible, although other measures usually are needed to control the hypercalcemia.

Summary

Hypercalcemia is one of the most common of metabolic disorders. Although there are many causes of hypercalcemia, primary hyperparathyroidism (PH) and malignancy account for the majority of the cases.

In the clinic, hyperparathyroidism is likely to be the cause, while malignancy is the likely cause in the hospital. The hypercalcemia of malignancy can occur with or without bone metastases. In both circumstances, the production by the tumor of parathyroid hormone-related protein (PTHrP) is the most common cause of the increased osteoclastic bone resorption that causes the hypercalcemia. Hypercalcemia can occur through one or a combination of three mechanisms:

- Increased bone resorption
- Increased gastrointestinal (GI) absorption of calcium
- Decreased renal calcium excretion.

The dominant clinical effects of significant hypercalcemia are on the nervous system, causing symptoms ranging from vague complaints to coma. Treatment of hypercalcemia of any cause can be effective, especially in the short term. The same three mechanisms that cause hypercalcemia are reversed to treat hypercalcemia:

- Inhibit bone resorption
- Decrease GI absorption of calcium
- Increase renal excretion of calcium.

A model regimen (Table 3.6) for treating hypercalcemia includes:

- Hydration
- Calciuresis
- Administration of inhibitors of bone resorption.

Saline is favored for the hydration and calciuresis unless there are cardiovascular contraindications. Calciuresis can be enhanced in the hydrated patient with loop diuretics. Salmon calcitonin (SCT) and the bisphosphonate pamidronate are used to inhibit bone resorption, with the former having early but mild effects and the latter having delayed but potent and sus-

TABLE 3.6 — TREATMENT OF HYPERCALCEMIA: MODEL REGIMEN
• *Hydration*: vigorous hydration with saline-containing fluids (eg, 5 to 10 L/d of isotonic – 1/2 normal saline in 24 hours). Monitor cardiovascular system. • *Calciuresis*: after adequate hydration, start furosemide (Lasix) (100 mg/2 hours) or ethacrynic acid (Edecrin) (40 mg/2 hours). Correct resulting hypokalemia and hypomagnesemia. • *Calcitonin* (Miacalcin): initiate early and at maximum dose (up to 8 U/Kg/q6h). • *Pamidronate* (Aredia): initiate early at one of several IV dose regimens (60 to 90 mg in 4 to 24 hours) • Collect appropriate samples for testing (eg, serum PTH) before treatment. • *Primary Disease*: direct treatment.

tained effects. Both should be used since they complement each other. Treatment of the hypercalcemia provides time for diagnosis and treatment of the underlying cause of the hypercalcemia.

SUGGESTED READING

Bosch X. Hypercalcemia due to endogenous overproduction of active vitamin D in identical twins with cat-scratch disease. *JAMA*. 1998;279:532-534.

Bosch X. Hypercalcemia due to endogenous overproduction of 1,25-dihydroxyvitamin D in Crohn's disease. *Gastroenterology*. 1998;114:1061-1065. Abstract.

Brincker H, Westin J, Abildgaard N, et al. Failure of oral pamidronate to reduce skeletal morbidity in multiple myeloma: a double-blind placebo-controlled trial. *Br J Haematol*. 1998;101: 280-286.

Collins MT, Skarulis MC, Bilezikian JP, Silverberg SJ, Spiegel AM, Marx SJ. Treatment of hypercalcemia secondary to parathyroid carcinoma with a novel calcimimetic agent. *J Clin Endocrinol Metab*. 1998;83:1083-1088.

Deftos LJ. Hypercalcemia: mechanisms, differential diagnosis, and remedies. *Postgrad Med.* 1996;100:119-121, 125-126.

Deftos LJ, Burton DW, Baird SM, Terkeltaub RA. Hypercalcemia and systemic lupus erythematosus. *Arthritis Rheum.* 1996;39:2066-2069.

Diel IJ, Solomayer E-F, Costa SD, et al. Reduction in new metastases in breast cancer with adjuvant clodronate treatment. *N Engl J Med.* 1998;339:357-363.

Dockrell DH, Poland GA. Hypercalcemia in a patient with hypoparathyroidism and *Nocardia asteroides* infection: a novel observation. *Mayo Clin Proc.* 1997;72:757-760.

Dusso AS, Kamimura S, Gallieni M, et al. γ-Interferon-induced resistance to 1,25-$(OH)_2$ D_3 in human monocytes and macrophages: a mechanism for the hypercalcemia of various granulomatoses. *J Clin Endocrinol Metab.* 1997;82:2222-2232.

Endo K, Katsumata K, Iguchi H, et al. Effect of combination treatment with a vitamin D analog (OCT) and a bisphosphonate (AHPrBP) in a nude mouse model of cancer-associated hypercalcemia. *J Bone Miner Res.* 1998;13:1378-1383.

Guise TA, Mundy GR. Cancer and bone. *Endocr Rev.* 1998;19:18-54.

Iguchi H, Miyagi C, Tomita K, et al. Hypercalcemia caused by ectopic production of parathyroid hormone in a patient with papillary adenocarcinoma of the thyroid gland. *J Clin Endocrinol Metab.* 1998;83:2653-2657.

Ishikawa S, Ozaki T, Kawai A, Inoue H, Doihara H. Hyperparathyroid crisis in a patient with a giant brown tumor of the iliac bone: a case report. *Hiroshima J Med Sci.* 1998;47:27-30.

Knecht TP, Behling CA, Burton DW, Glass CK, Deftos LJ. The humoral hypercalcemia of benignancy. A newly appreciated syndrome. *Am J Clin Pathol.* 1996;105:487-492.

Lee JC, Catanzaro A, Parthemore JG, Roach B, Deftos LJ. Hypercalcemia in desseminated coccidioidomycosis. *N Engl J Med.* 1977; 297:431-433.

Leifsson BG, Ahren B. Serum calcium and survival in a large health screening program. *J Clin Endocrinol Metab.* 1996;81:2149-2153.

Malik GM. High serum calcium in human brucellosis: a case-control study. *Am J Trop Med Hyg*. 1998;59:397-398.

Mundy GR, Guise TA. Hypercalcemia of malignancy. *Am J Med*. 1997;103:134-145.

Pearce SHS. Endocrinology: calcium homeostasis and disorders of the calcium-sensing receptor. *J R Coll Physicians Lond*. 1998;32: 10-13.

Philbrick WM. Parathyroid hormone-related protein is a developmental regulatory molecule. *Eur J Oral Sci*. 1998;106:32-37.

Rosen HN, Lim M, Garber J, et al. The effect of PTH antagonist BIM-44002 on serum calcium and PTH levels in hypercalcemic hyperparathyroid patients. *Calcif Tissue Int*. 1997;61:455-459.

Saarto T, Blomqvist C, Risteli J, Risteli L, Sarna S, Elomaa I. Aminoterminal propeptide of type I procollagen (PINP) correlates to bone loss and predicts the efficacy of antiresorptive therapy in pre- and post-menopausal non-metastatic breast cancer patients. *Br J Cancer*. 1998;78:240-245.

Sachmechi I, Shah G, Rezainadimi L, Blaustein DA, Rosner F. Misleading acute hypercalcemia due to hyperlioidemia: a method-dependent error. *Endocr Pract*. 1997;3:293-296.

Sakoulas G, Tritos NA, Lally M, Wanke C, Hartzband P. Hypercalcemia in an AIDS patient treated with growth hormone. *AIDS*. 1997;11:1353-1356.

Silva LCS, Ferrari TCA. Hypercalcaemia and paracoccidioidomycosis. *Transactions R Soc Trop Med Hyg*. 1998;92:187.

Silverberg SJ, Bone HG III, Marriott TB, et al. Short-term inhibition of parathyroid hormone secretion by a calcium-receptor agonist in patients with primary hyperparathyroidism. *N Eng J Med*. 1997;337:1506-1510.

Stewart AF, Horst R, Deftos LJ, Cadman EC, Lang R, Broadus AE. Biochemical evaluation of patients with cancer-associated hypercalcemia: evidence for humoral and nonhumoral groups. *N Engl J Med*. 1980;303:1377-1383.

Wysolmerski JJ, Stewart AF. The physiology of parathyroid hormone-related protein: an emerging role as a developmental factor. *Annu Rev Physiol*. 1998;60:431-460.

Young G, Shende A. Use of pamidronate in the management of acute cancer-related hypercalcemia in children. *Med Pediatr Oncol.* 1998;30:117-121.

Zeimer HJ, Greenaway TM, Slavin J, et al. Parathyroid-hormone-related protein in sarcoidosis. *Am J Pathol.* 1998;152:17-21.

4 Primary Hyperparathyroidism

Introduction

4

Primary hyperparathyroidism (PH) is one of the most common causes of hypercalcemia. Following the introduction of multichannel testing that included calcium measurement, the incidence of PH increased 25- to 30-fold in the 1970s to approximately 150 to 200 cases/million population/year. Recently, the incidence has decreased toward precalcium screening levels for unclear reasons. Nevertheless, in hospitals, it is second only to malignancy as a cause of hypercalcemia; in the outpatient setting, PH is the most common cause. Whereas malignancy dominates the clinical picture when it causes hypercalcemia, in PH the physician in the clinic is often faced with the challenge of establishing the cause of an elevated calcium that was part of multichannel screening rather than a specific request, even though the advent of managed care has reduced the use of such screening tests. However, the diagnosis of hyperparathyroidism is relatively easy to establish under these circumstances. The key biochemical features of PH are:

- Elevated serum calcium
- Decreased fasting serum phosphorus
- Increased serum parathyroid hormone (PTH).

The history can be helpful in distinguishing PH from the hypercalcemia of malignancy. In the former, there is often a long (weeks to months) history of vague symptomatology and a record of an overlooked elevated calcium level in the past. Next to abnormal thyroid tests, physicians most tend to ignore elevated serum calcium levels that appear on a multichannel screening test. Once the diagnosis is made, therapy can be addressed.

Etiology and Pathogenesis

Primary hyperparathryoidism is a neoplastic disorder of the parathyroid glands characterized by increased secretion of PTH. PH is usually due to a single adenoma of the gland (85%), but can be due to the multiple adenomas; hyperplasia of all glands (15%), usually four; or carcinoma (< 1%). Hyperplasia can occur by itself or as a part of multiple endocrine neoplasia (MEN) MEN types 1 and 2, especially MEN type 2A, and familial hyperparathyroidism (FH). FH can also occur by itself, referred to as familial isolated hyperparathyroidism (FIHP), or in association with jaw tumors and mapping to chromosome 1, referred to as the hyperparathyroidism-jaw tumor syndrome (HPT-JT). There is an increased incidence of multiple parathyroid adenomas and cancers in these two syndromes. Ectopic parathyroid glands occur in approximately 5% of patients with PH.

PH usually occurs between ages 40 to 65, and is 2 to 3 times more common in females than males. The approximate incidence of one per 1000 seems to be declining. Transient hyperparathyroidism can occur in the newborn of a hypocalcemic mother. There are several candidate oncogenes for PH, especially cyclin D1/PRAD1 gene and the MEN type 1 tumor suppressor gene (Chapter 6, *Multiple Endocrine Neoplasias*), but none has been conclusively identified as causative.

PH is now usually diagnosed in a nonsymptomatic patient found to be incidentally hypercalcemic by a multichannel screening test. Such presumably early diagnosis finds the patient without the classical manifestations of the disease. These patients are characterized as having asymptomatic hyperparathyroidism, which actually means no overt signs or symptoms of the disease. In general, a serum calcium under 12 mg/dL does not cause symptoms. However, in some countries and regions where medical care is poor and where there is vitamin D deficiency, the classical manifestations of PH still appear. The classical manifestations reflect the biological actions of the PTH excess. These symptoms can be subsumed under the clinical rubrics of "stones, bones, abdominal groans, and psychic moans" (Table 4.1).

Clinical Features

■ Renal

Hypercalciuria and nephrolithiasis remain among the most common manifestations of PH, but occur in only 15% of patients; conversely, 7% of calcium-stone formers have PH. The calcium-containing stones are related to hypercalciuria, but other factors seem to play a role. Nephrocalcinosis is rare, but the increased ambient calcium can interfere with renal tubular function and cause polyuria and polydypsia. Although rare, there can be a gradual loss of renal function, which can be stabilized and sometimes reversed by successful treatment (ie, parathyroidectomy [PTX]).

■ Skeletal

The skeletal manifestations of PH are relatively mild in the contemporary patient, with osteitis fibrosa cystica, brown tumors, cystic lesions, and pathological fractures now being rare. Nevertheless, increased bone turnover is still seen as reflected by increased markers of bone formation, such as total and bone-specific alkaline phosphatase, and bone resorption, such as collagen-related measurements (see Chapter 1, *Skeletal Metabolism*, and Tables 1.9 and 1.10). While bone resorption exceeds formation in cortical

TABLE 4.1 — SIGNS AND SYMPTOMS OF PRIMARY HYPERPARATHYROIDISM
Renal • Nephrolithiasis ("Stones") • Hypercalciuria • Nephrocalcinosis • Renal tubular defects • Polyuria/polydypsia • Renal failure • Hypertension
Skeletal ("Bones") • Subperiosteal resorption • Osteitis fibrosa cystica • Bone cysts • Brown tumors • Osteoclastomas • Demineralization • Decreased bone mineral density • Pathologic fractures
Gastrointestinal ("Groans) • Nausea and vomiting • Constipation • Peptic ulcers • Pancreatitis
Nervous System ("Moans") • Depression • Lethargy and fatigue • Memory loss and confusion • Depressed reflexes • Neuroses • Psychoses • Coma
Other • Hypertension • Anemia • Itching • Calcific keratitis and conjunctivitis • Muscle weakness

bone, cancellous bone volume is usually maintained and trabecular structure can be preserved. Thus, the radius, primarily cortical bone at its distal third site, is affected more than the hip, and the spine is the least affected. Consequently, forearm densitometry (at the distal third of the radius) is the most sensitive site for detecting skeletal involvement in PH, while measurements at the spine may be normal. Overall bone mineral density (BMD) is reduced, but can return toward normal after curative PTX with substantial increases (over 10%), especially at the spine, occurring within a few years after surgery and sustained for up to 10 years.

Bone densitometry can be used to demonstrate the subtle changes in the skeleton that take place in PH, since x-rays become abnormal only when there is 30% to 50% demineralization. In fact, x-rays may not be indicated if there are no skeletal symptoms and the alkaline phosphatase is normal. However, x-rays can still be used to demonstrate the abnormal bone resorption that now rarely occurs in the subperiosteum of the phalanges and distal clavicles, the skull ("salt and pepper"), and the lamina dura of teeth. Chondrocalcinosis and pseudogout occur in up to 5% of patients, the latter also occurring post-PTX. The now rare bone cysts and brown tumors will usually be radiolucent.

■ Gastrointestinal

Nausea and vomiting can complicate significant hyperparathyroidism and thus contribute to the dehydration. Constipation seems to be a common feature, while an association with peptic ulcer disease (in the absence of a Zollinger-Ellison [ZE] syndrome) is tenuous. However, constipation is too common a symptom in the general population to be diagnostically useful. Pancreatitis is also a rare complication, due presumably to calcium deposits in pancreatic ducts.

■ Nervous System

Severe hypercalcemia can lead to depressed nervous system function that can result in coma. Many patients with other than severe disease commonly complain of lethargy, depression, and memory loss. However, a causal relationship of the PH for these nonspecific symptoms has not been established. Mental obtundation up to coma can occur in the severely hypercalcemic patient.

■ Other Complications

A variety of complications have been attributed to PH, notably anemia and hypertension. Some of these have been related to a presumed toxic effect of the high levels of PTH. However, these relationships are not causally established. Itching may occur when the calcium is exceedingly high. Calcific keratis and muscle weakness are rare complications of PH.

Laboratory Findings

Hypercalcemia and increased serum PTH are the signal manifestations of PH, although levels of both can fluctuate close to the upper limits of normal early in the disease as it waxes and wanes. Hypophosphatemia (25%) results from the phosphaturic effect of PTH, but to be of diagnostic value, this measurement (unlike calcium) must be made in the fasting state because of post-prandial fluctuations in serum phosphorus. Total serum calcium measurement will usually suffice, but when there are changes in serum proteins (especially albumin) that can effect this measurement, ionized calcium can be helpful in establishing true hypercalcemia. The renal actions of PTH on chloride and bicarbonate may cause a mild hypercholoremic acidosis. Urinary calcium is often (40%) elevated, and this test can be used to help exclude the diagnosis of familial hypocalcuric hypercalcemia

(FHH). However, urinary calcium can be as high as 250 mg/24 hr in FHH. 1,25-Dihydroxyvitamin D (1,25-D) can be increased (30%) because of the stimulatory effect of PTH on renal 1α-hydroxylase (see Chapter 2, *Calcium Metabolism*), and bone markers can be elevated (Chapter 1, *Skeletal Metabolism*).

More than one measurement of serum calcium and sometimes PTH should be made for a combination of technical and biological reasons. Since PH waxes and wanes, especially during its early course, a single measurement might miss the wax. Serum PTH is now most commonly measured by a two-site assay that measures the intact molecule. However, it has been recently reported that PTH fragments may also be measured by such assays. Nevertheless, this is the most widely used test, even though well characterized mid-molecule assays are equally accurate and precise in diagnosis. Amino-terminal tests (N-PTH) are not widely used. The upper limit of normal for intact PTH is usually 65 pg/mL. However, after the age of about 45 years, even intact PTH seems to increase in normal subjects, perhaps due to decline in renal function. Thus, in the younger hypercalcemic patient, an intact PTH above the mid-forties level should be suspicious for PH.

There are converse considerations with regard to serum calcium measurements. Most laboratories place the upper limit of normal at 10.2 to 10.5 mg/dL. However, serum calcium slowly declines in the aging individual. Thus, after approximately age 50, a serum calcium above 10 mg/dL should also be regarded with the appropriate clinical suspicion.

The relationship between serum calcium and PTH measurement should also be considered. Since even in PH calcium feeds back on the secretion of PTH, a serum PTH at the upper limits of normal may be considered as increased in the face of a serum calcium above normal.

Treatment

■ Surgery

Surgery is the treatment of choice and is usually curative. A consensus development conference of the National Institutes of Health (NIH) established criteria for surgery to include any of the following. (However, it should be noted that many experts do not agree with these recommendations, notably the serum and urine cut-offs):

- Serum calcium > 11.4 to 12 mg/dL
- Urinary calcium > 400 mg/24 hrs in males and 350 mg/24 hrs in females
- Bone density (cortical) Z score < 2.0, especially at the distal third of the radius
- Substantial decrease in creatinine clearance
- Age < 50
- Overt manifestations such as:
  - Nephrolithiasis
  - X-ray evidence of disease
  - Neuromuscular disease
  - Symptoms of disease
- A previous episode of health-threatening hypercalcemia
- Unlikley success of medical management.

Approximately half of patients with PH meet one of these criteria, and of them about half eventually undergo surgery. Some of the surgical choices are summarized in Table 4.2. The usual surgery is removal of a single adenoma with or without further exploration. For hyperplasia, all glands should be identified, with special attention to the increased likelihood of aberrant tissue. All hyperplastic tissue can be removed, but some surgeons attempt to leave behind one half of one gland. The former patients can be treated with calcium and vitamin D, the latter monitored for PTH sufficiency.

TABLE 4.2 — SURGERY OF PRIMARY HYPERPARATHYROIDISM
• Unilateral parathyroidectomy and evaluation of ipsilateral gland • Bilateral exploration • Removal of adenoma and biopsy of one or two other glands • Removal of any enlarged gland and biopsy of all others • Subtotal parathyroidectomy • Total parathyroidectomy and autotransplantation

Implantation of parathyroid tissue in the forearm or strap muscles should be undertaken with caution in PH, since the transplanted tissue can resume its autonomous growth and return the patient to the hyperparathyroid state.

Although most patients will meet the criteria for surgery, a majority of them will be asymptomatic. The final decision for surgery accommodates not only these criteria but personal considerations of the patient and of the surgeon's skill as well. For example, a patient may decline surgery for a variety of substantial or superficial reasons, or there may be a contraindication such as a recent heart attack. When surgery is declined, the patient should be followed every 6 to 12 months with tests that assess disease status, such as measurements of serum calcium and PTH, urinary calcium, tests of renal function, and every year or so with bone densitometry.

When surgery is chosen, referral to an experienced parathyroid surgeon is critical, where the success rate is 95% for a single adenoma and 80% for multiple gland surgery. However, surgeons without extensive parathyroid experience may not be as successful (Table 4.3). The major complications of surgery include treatment failure, hypoparathyroidism (especially when surgery is extensive), and recurrent

TABLE 4.3 — RESULTS OF PARATHYROIDECTOMY	
Cure	> 90%
Mortality	< 0.2%
Vocal cord paralysis	~ 0.2%
Wound repair	< 2%
Respiratory problems	< 1.5%
Hypocalcemia (transient)	~ 10%
Hyypoparathyroidism	~ 2%

laryngeal nerve damage. Their incidence is less than 1% for the experienced surgeon. Ultrasonography-guided ablation of the adenoma is still experimental and hazardous. Successful treatment will cause the biochemical and skeletal abnormalities to stabilize or reverse, and there is usually improvement in symptoms. Decreased bone mass can revert toward normal, especially the spine.

Following the patient with mild asymptomatic hypercalcemia is another option. Long-term studies indicate that the hypercalcemia and BMD in such patients may not change over a decade. Some studies report an increased mortality in the chronically hypercalcemic patient, while others do not. The expense of long-term follow-up may exceed that of curative surgery. Thus, decisions should be tailored to the individual patient.

■ Pre- and Intraoperative Surgical Localization

Many procedures have come and gone for preoperative localization of the abnormal parathyroid tissue (Table 4.4). Among those in current use are ultrasound (US), computer-assisted tomography (CAT), magnetic resonance imaging (MRI), and ^{99m}Tc-sestamibi scanning, with sensitivities of 60% to 80%.

TABLE 4.4 — PREOPERATIVE LOCALIZATION OF ABNORMAL PARATHYROID TISSUE
• Physical examination • Thyroid scan • Cervical esophagram • Venous sampling: neck massage • Angiography • Thermography • Technetium/thallium scanning • Lymphography • Computer-assisted tomography (CAT) • Ultrasound • ^{99m}Tc-Sestamibi scan

4

This sensitivity may be improved by the use of adjunctive computerized tomography. Angiography and venous sampling for PTH are saved for only the most difficult cases, usually involving reoperation.

Of the several procedures available for the preoperative localization of the abnormal parathyroid tissue, each new method arrives with fanfare, but experience reveals the presence of false-positives and false-negatives with most procedures. Although many surgeons request these localizing procedures, with ^{99m}Tc-sestamibi being the latest favorite, there is little evidence that they increase success rates or decrease duration and morbidity of surgery, although they may have some such value in reoperations. Thus, locating an experienced surgeon remains the best preoperative localization procedure, and endocrinologists regard the skilled and experienced surgeon's fingers as the best localizing tools.

Intraoperative PTH measurements are being introduced to monitor surgery, but they are not widely available. Although such measurements of PTH have been used to evaluate the success of surgery, the best intraoperative localization tools are also the fingers of an experienced surgeon.

■ Medical Treatment of Primary Hyperparathyroidism

Medical therapy for PH is generally ineffective. In postmenopausal women, high doses of estrogens (eg, 0.625 mg to 1.25 mg of Premarin) may lower blood calcium by opposing the effects of PTH, especially on bone. However, serum PTH and 1,25-D remain elevated. Progestins have similar effects but adversely influence blood lipids. Raloxifene may be uniquely useful for this indication (see Chapter 10, *Osteoporosis*). Calcitonin (CT) may decrease serum calcium, but limited experience with it and the bisphosphonates has not shown them to be effective long term, and the serum PTH can increase further. Oral phosphate may lower calcium, but at the risk of extraskeletal calcification and a further increase in PTH; intravenous phosphorus therapy is no longer used because it exacerbates these risks.

Calcimimetic drugs that block the calcium sensor on the parathyroid glands are under experimental evaluation and show considerable promise (Chapter 2, *Calcium Metabolism*). Their hypercalciuric effect is of some concern. Diets high in calcium and vitamin D should be avoided, but low calcium diets are unpalatable and may lead to increased serum PTH. Thiazide diuretics, which decrease renal calcium excretion, should be avoided. Loop diuretics, useful in the acute treatment of hypercalcemia because of their calciuretic action, have no role in sustained medical management.

Postparathyroidectomy Hypocalcemia

Postoperative hypocalcemia can be a confusing and dramatic complication of PTX. It is more likely to occur when there is extensive surgery and manipulation of the parathyroid glands. Preoperative evidence of bone disease, such as elevated bone markers and abnormal x-rays, increase its likelihood. The hypocalcemia can be due to two basic mechanisms: hypoparathyroidism or recalcification of demineralized bone (the "hungry bone" syndrome). Table 4.5 summarizes their pathogenetic mechanisms. Table 4.6 provides the bases for making the distinction between the two conditions.

TABLE 4.5 — MECHANISMS OF POSTPARATHYROIDECTOMY HYPOCALCEMIA
Decreased Parathyroid Hormone Secretion (Hypoparathyroidism) • Removal of hyperfunctioning tissue – Adenoma – Hyperplasia • Suppression of remaining tissue – Hypercalcemia – Trauma at surgery
Recalcification of Demineralized Bone • "Hungry bones"

Treatment of post-PTX hypocalcemia can be considered in three phases:

- Emergency
- Intermediate
- Maintenance.

Specific protocols for each of these phases are summarized in Table 4.7.

Special attention must be given to the serum magnesium, since it can make the hypocalcemia difficult or refractory to treatment if low. Because serum magnesium is not a sensitive and specific index of magnesium status, a therapeutic trial of magnesium may be prudent even if the serum level is normal. Table

TABLE 4.6 — POSTPARATHYROIDECTOMY HYPOCALCEMIA: DIFFERENTIAL DIAGNOSIS

	Hypoparathyroidism	"Hungry Bones"*
Preoperative evidence for bone disease	No	Yes
Extensive manipulation of parathyroid glands	Yes	No
Serum phosphate	↑	↓
Urinary calcium	↑	↓
Parathyroid hormone	↓	↑
Calcium requirements	< 500 mg/24 hrs Chronic Calcium and vitamin D	> 500 mg/24 hrs Acute Calcium only

* The recalcification of demineralized bone — "hungry bones" syndrome.

TABLE 4.7 — TREATMENT OF POSTPARATHYROIDECTOMY HYPOCALCEMIA	
Emergency	• 20 to 30 mL of 10% calcium (9 mg/mL) gluconate (100 mL D_5W) in 10 to 15 minutes IV • For mg < 0.8 mEq/L, 1 to 2 g of 10% $MgSO_4$ in 10 to 15 minutes IV or 2 mL IM (50%)
Intermediate	• Titrate serum calcium toward normal every 6 to 12 hours with 500 mg calcium in 1000 mL D_5W • Titrate serum magnesium to high normal with 50 to 100 mEq/24 hours or 1 g IM every 4 to 6 hours (care in renal failure)
Maintenance	• Calcium (many preparations, Table 5.5): – Vitamin D after 2 weeks – Monitor blood and urinary calcium – Phosphate depletion – Monitor serum magnesium • Magnesium: diarrheagenic
Abbreviations: D_5W, dextrose 5% in water (solution); IV, intravenous; IM, intramuscular; $MgSO_4$, magnesium sulfate.	

4.8 provides guidelines for evaluating the effectiveness of magnesium therapy. The various preparations of vitamin D and calcium available for the treatment of hypoparathyroidism are summarized in Table 5.4 (see Chapter 5, *Hypocalcemia and Hypoparathyroidism*, and Chapter 7, *Osteomalacia and Rickets*).

Summary

Primary hyperparathyroidism (PH) is the most common endocrine cause of hypercalcemia, and the

TABLE 4.8 — END-POINTS IN THE TREATMENT OF POSTPARATHYROIDECTOMY HYPOCALCEMIA
Neurological Signs and Symptoms • Tetany and paresthesias • Chvostek's and Trousseau's signs
Check Magnesium > 0.8 to 1.0 mEq/L
Low Normal Calcium > 7.5 mg/dL • Enhances parathyroid hormone secretion (conserves phosphorus) • Permits detection of eucalcemia
Abbreviations: PTH, parathyroid hormone.

most common cause of PH is a single adenoma of the parathyroid glands. There are about 4 million cases of PH world wide. Rather than presenting to the physician with symptoms, the patient with PH is more likely to be asymptomatic and seen because of an elevated serum calcium detected by a multi-channel screening. When there are clinical manifestations of PH, they involve the:

- Skeleton
- Gastrointestinal tract
- Kidney
- Central nervous system.

Primary hyperparathyroidism can occur by itself as a sporadic or familial disease or as part of the multiple endocrine neoplasia (MEN) or familial hyperparathyroid (FH) syndromes. Females between the ages of 40 to 65 are the most commonly involved group. PH can usually be distinguished from the other common cause of hypercalcemia, malignancy, by a careful history. An elevated serum calcium and decreased fasting serum phosphorus support the diagnosis of PH. And an elevated serum parathyroid hormone (PTH) establishes the diagnosis. By contrast, in the hypercalcemia of malignancy the serum phosphorus does not have a consistent pattern, and the serum PTH is usually suppressed by the hypercalcemia.

When PH seems likely, the evaluation can often be completed by confirmation with an elevated serum PTH. If the diagnosis is not clear, then it may be prudent and ultimately cost effective to order the other tests discussed in the differential diagnosis of hypercalcemia (see Chapter 3, *Hypercalcemia*, especially Table 3.3).

4

Parathyroid gland localizing procedures, with ^{99m}Tc-sestamibi scan being now most popular, should not be used to make the diagnosis of PH, but may be used to localize the offending tissue after the diagnosis has been made and in anticipation of surgery. However, the skilled and experienced parathyroid surgeon is the best localizer of abnormal parathyroid glands.

Parathyroidectomy (PTX) is the treatment of PH and is usually curative. The experienced surgeon has few complications. Patients with mild hypercalcemia and without any complications of PH may be followed, especially if there are any contraindications to surgery. Medical treatment is generally not effective, but new drugs, especially the bisphosphonates and calcimimetics, are under study.

SUGGESTED READING

Apostolopoulos DJ, Houstoulaki E, Giannakenas C, et al. Technetium-99m-tetrofosmin for parathyroid scintigraphy: comparison to thallium-technetium scanning. *J Nucl Med.* 1998;39:1433-1441.

Collins MT, Skarulis MC, Bilezikian JP, Silverberg SJ, Spiegel AM, Marx SJ. Treatment of hypercalcemia secondary to parathyroid carcinoma with a novel calcimimetic agent. *J Clin Endocrinol Metab.* 1998;83:1082-1088.

Deftos LJ. Markers of bone turnover in primary hyperparathyroidism. In: Bilezikian JP, Marcus R, Levine MA, eds. *The Parathyroids: Basic and Clinical Concepts*. New York, NY: Raven Press Ltd; 1994:485-492.

Deftos LJ, Parthemore JG, Stabile BE. Management of primary hyperparathyroidism. *Annu Rev Med*. 1993;44:19-26.

Delbridge LW, Younes NA, Guinea AI, Reeve TS, Clifton-Bligh P, Robinson BG. Surgery for primary hyperparathyroidism 1962-1996: indications and outcomes. *Med J Aust*. 1998;168:153-156.

Glendenning P, Gutteridge DH, Retallack RW, Stuckey BGA, Kermode DG, Kent GN. High prevalence of normal total calcium and intact PTH in 60 patients with proven primary hyperparathyroidism: a challenge to current diagnostic criteria. *Aust NZ J Med*. 1998;28:173-178. Abstract.

Guise TA, Yin JJ, Taylor SD, et al. Evidence for a causal role of parathyroid hormone-related protein in the pathogenesis of human breast cancer-mediated osteolysis. *J Clin Invest*. 1996;98:1544-1549.

Hedback G, Oden A. Increased risk of death from primary hyperparathyroidism: an update. *Eur J Clin Invest*. 1998;28:271-276.

Hines OJ, Gordon HE. Secondary hyperparathyroidism in a patient with eight parathyroid glands. *J Clin Endocrinol Metab*. 1998; 83:2384-2386.

Ishibashi M, Nishida H, Hiromatsu Y, Kojima K, Tabuchi E, Hayabuchi N. Comparison of technetium-99m-MIBI, technetium-99m-tetrofosmin, ultrasound and MRI for localization of abnormal parathyroid glands. *J Nucl Med*. 1998;39:320-324.

Khafif A, Halperin D, Marshak G. Ethanol injection to parathyroid tissue: indications and limitations. *Ear Nose Throat J*. 1998;77:538-540.

Kulak CA, Bandeira C, Voss D, et al. Marked improvement in bone mass after parathyroidectomy in osteitis fibrosa cystica. *J Clin Endocrinol Metab*. 1998;83:732-735.

Loh KC, Duh QY, Shoback D, Gee I, Siperstein A, Clark OH. Clinical profile of primary hyperparathyroidism in adolescents and young adults. *Clin Endocrinol*. 1998;48:435-443.

Nemeth EF, Steffey ME, Hammerland LG, et al. Calcimimetics with potent and selective activity on the parathyroid calcium receptor. *Proc Natl Acad Sci U S A*. 1998;95:4040-4045.

Nussbaum SR, Gaz RD, Arnold A. Hypercalcemia and ectopic secretion of parathyroid hormone by an ovarian carcinoma with rearrangement of the gene for parathyroid hormone. *N Engl J Med*. 1990;323:1324-1328.

Ott SM. Calcimimetics–new drugs with the potential to control hyperparathyroidism. *J Clin Endocrinol Metab*. 1998;83:1080-1082. Editorial.

Palanisamy N, Imanishi Y, Rao PH, Tahara H, Chaganti RS, Arnold A. Novel chromosomal abnormalities identified by comparative genomic hybridization in parathyroid adenomas. *J Clin Endocrinol Metab*. 1998;83:1766-1770.

Proceedings of the NIH Consensus Development Conference on diagnosis and management of asymptomatic primary hyperparathyroidism. Bethesda, Maryland, October 29-31, 1990. *J Bone Min Res*. 1991;6(suppl 2):S1-S166.

Rosen HN, Lim M, Garber J, et al. The effect of PTH antagonist BIM-44002 on serum calcium and PTH levels in hypercalcemic hyperparathyroid patients. *Calcif Tissue Int*. 1997;61:455-459.

Sadler GP, Jones DL, Morgan JM, Neonakis E, Woodhead JS, Wheeler MH. Role of octreotide on release of intact 1-84 parathyroid hormone from human parathyroid cells. *Br J Surg*. 1998;85:1133-1137.

Seymour JR, Gagel RF, Hagemeister FB, Dimopoulos MA, Cabanillas F. Calcitriol production in hypercalcemia and normocalcemia patients with non-Hodgkin lymphoma. *Ann Intern Med*. 1994;121:633-640.

Sherrard DJ. Calcimimetics in action. *Kidney International*. 1997;53:510-511. Editorial.

Silverberg SJ, Bilezikian JP. Primary hyperparathyroidism: still evolving? *J Bone Miner Res*. 1997;12:856-862.

Silverberg SJ, Bone HG III, Marriott TB, et al. Short-term inhibition of parathyroid hormone secretion by a calcium-receptor agonist in patients with primary hyperparathyroidism. *N Engl J Med*. 1997;337:1506-1510.

Simon GH, Lakritz P, Freeman LM. Partial ablation of mediastinal parathyroid adenoma by angiography: documentation with Tc-99m sestamibi scintigraphy. *Clin Nucl Med.* 1998;23:174-176.

Sosa JA, Powe NR, Levine MA, Udelsman R, Zeiger MA. Profile of a clinical practice: thresholds for surgery and surgical outcomes for patients with primary hyperparathyroidism: a national survey of endocrine surgeons. *J Clin Endocrinol Metab.* 1998;83:2658-2665.

Stewart AF, Horst R, Deftos LJ, Cadman EC, Lang R, Broadus AE. Biochemical evaluation of patients with cancer-associated hypercalcemia: evidence for humoral and nonhumoral groups. *N Engl J Med.* 1980;303:1377-1383.

Tahara H, Smith AP, Gaz RD, Zariwala M, Xiong Y, Arnold A. Parathyroid tumor suppressor on 1p: analysis of the p18 cyclin-dependent kinase inhibitor gene as a candidate. *J Bone Miner Res.* 1997;12:1330-1334.

Teh BT, Farnebo F, Twiff S, et al. Familial isolated hyperparathyroidism maps to the hyperparathyroidism-jaw tumor locus in 1q21-q32 in a subset of families. *J Clin Endocrinol Metab.* 1998;83: 2114-2120.

Vincente Torregrosa J, Rosa Palomar M, Pons F, et al. Has double-phase MIBI scintigraphy usefulness in the diagnosis of hyperparathyroidism? *Nephrol Dial Transplant.* 1998;13:37-40.

Vogel LM, Lucas R, Czako P. Unilateral parathyroid exploration. *Am Surg.* 1998;64:693-696.

Wermers RA, Khosla S, Atkinson EJ, et al. Survival after the diagnosis of hyperparathyroidism: a population-based study. *Am J Med.* 1998;104:115-122.

Wermers RA, Khosla S, Atkinson EJ, Hodgson SF, O'Fallon WM, Melton LJ III. The rise and fall of primary hyperparathyroidism: a population-based study in Rochester, Minnesota, 1965-1992. *Ann Intern Med.* 1997;126:433-440.

Westra WH, Pritchett DD, Udelsman R. Intraoperative confirmation of parathyroid tissue during parathyroid exploration: a retrospective evaluation of the frozen section. *Am J Surg Pathol.* 1998; 22:538-544.

5 Hypocalcemia and Hypoparathyroidism

Introduction

The most common cause of hypocalcemia is renal failure (see Chapter 9, *Renal Osteodystrophy*), and the other etiologies account for only a small minority of patients with hypocalcemia. The clinical manifestations of hypocalcemia are due to the increased neuromuscular excitability that accompanies a low serum calcium. Alkalosis will aggravate hypocalcemia by causing a further decline in ionized calcium. Patients complain of paresthesias and tingling in the extremities and about the mouth. Chvostek's and Trousseau's signs can be elicited. Tetany and convulsions are the frankest manifestations of the increased neuromuscular irritability.

The hypocalcemic symptoms are related to the absolute level of serum calcium and to the changes in serum calcium. The major signs and symptoms of hypocalcemia are summarized in Table 5.1. The major causes of hypocalcemia are summarized in Table 5.2. Measuring the serum phosphorus, parathyroid hormone (PTH), and assessing renal function can usually identify the cause of the hypocalcemia. Artifactual causes of hypocalcemia should be ruled out. The most common artifactual cause is hypoalbuminemia, such as occurs in liver disease. The total serum calcium can be confirmed by an ionized calcium measurement or by correcting the total serum calcium according to the following formula: serum calcium + (0.8) (4-serum albumin).

TABLE 5.1 — SIGNS AND SYMPTOMS OF HYPOCALCEMIA
Neuromuscular • Paresthesias: mouth and extremities • Muscle spasms • Laryngeal strider • Seizures • Cardiac arrythmias • Coma • Chvostek's sign • Trousseau's sign: main d'accoucheur • Tetany – Clinical – Latent • Pseudotumor cerebri • Papilledema
Cardiac • Arrythmia • Electrocardiogram (EKG) abnormalities • Hypotension • Failure
Other • Cataracts: subcapsular, punctate • Extraskeletal calcification and ossification – Basal ganglia – Ligamentous • Clotting abnormalities • Psoriasis (von Zumbusch) • Alopecia • Xeroderma
Surgical Neck Scar

TABLE 5.2 — DIFFERENTIAL DIAGNOSIS OF HYPOCALCEMIA
• Renal failure • Hypoparathyroidism (Table 5.3) • Magnesium deficiency • Pancreatitis • Toxic shock syndrome • Osteoblastic metastases and "hungry bones" • Hyperphosphatemia • Treatment of hypercalcemia • Hypoproteinemia (hypoalbuminemia) • Calcium binding (chelation): – Citrated blood – Foscarnet – Phosphate administration – EDTA-containing contrast agents • Osteomalacia • Prematurity • Malabsorption • Vitamin D deficiency (Table7.1) • Blomstrand's chondrodysplasia • Acute adrenal insufficiency
Abbreviations: EDTA, ethylenediamine-tetraacetic acid.

5

Renal Failure

The pathogenesis of the hypocalcemia of renal failure is complex and discussed in detail in Chapter 9, *Renal Osteodystrophy*. The two primary causes are:

- Increased serum phosphorus
- Decreased renal production of 1,25-dihydroxy-vitamin D (1,25-D).

The former causes hypocalcemia by complexing with serum calcium and depositing it into bone and other tissues. The latter causes hypocalcemia by decreasing the gastrointestinal (GI) absorption of calcium. Both mechanisms initiate the complex pathways illustrated in Figure 9.1. Treatment is directed at:

- Correcting the renal failure
- Reducing serum phosphorus
- Increasing serum calcium by:
  - Adequate dialysis calcium concentration
  - Dietary calcium and vitamin D administration.

The preferred form of vitamin D is calcitriol, which can be given orally or parenterally.

Hypoparathyroidism

■ Pathogenesis

There are several causes of hypoparathyroidism, as summarized in Table 5.3. Neck surgery that removes or destroys the parathyroid glands is the most common cause of hypoparathyroidism. The most common neck surgeries causing hypoparathyroidism are cancer surgery and thyroidectomy, with inadvertent removal or destruction of parathyroid tissue, and parathyroidectomy (PTX), especially aggressively undertaken for hyperplasia.

All of the glands, usually four, have to fail in order for hypoparathyroidism to ensue. This can happen permanently following extensive surgery or it can happen transiently following the removal of adenoma whose secretion of PTH and resulting hypercalcemia has suppressed the other glands. The parathyroid glands usually recover within 1 week of surgery and after several weeks recovery is rare. In addition to this temporary-permanent dichotomy, surgical hypoparathyroidism can be acute or delayed in onset. The latter occurs when the glands are injured at surgery and the effects of the trauma progress. Postsurgical hypoparathyroidism can occur within hours after surgery or after the passage of time, usually weeks but sometimes months, when glands injured at surgery ultimately become nonfunctioning.

TABLE 5.3 — CAUSES OF HYPOPARATHYROIDISM
• Postoperative (acute and chronic): – Parathyroidectomy – Thyroidectomy – Cancer surgery: laryngeal, thyroid • Idiopathic: – Isolated – Associated with polyendocrine failure (Table 5.6) • Functional (decreased PTH secretion): – Magnesium deficiency – Activating mutation of the calcium sensor – Newborn of mother with PH • DiGeorge syndrome: aplasia/dysgenesis of the parathyroids and thymus (third and fourth pharyngeal pouches) • Familial hypoparathyroidism: – Calcium-sensor (CSR) mutations (activating) • Infiltrative: – Iron deposits: hemochemotosis/siderosis – Copper deposits: Wilson disease – Aluminum deposits: renal disease, aluminum exposure – Carcinoma – Amyloidosis – Sarcoid • Radiation to neck • Resistance to parathyroid hormone: – Pseudohypoparathyroidism (Table 5.7) – Renal failure – Magnesium deficiency

Tetany can occur 1 to 2 days postoperatively. Approximately half of patients with postoperative tetany do not require long-term therapy. Post-PTX hypocalcemia can be exaggerated by the "hungry bones" syndrome (Table 4.5). This occurs through the avid uptake of calcium by skeleton that has been demineralized by the actions of the high concentrations of PTH during the hyperparathyroidism. Removal of the abnormal parathyroid tissue reverses the increased bone resorption and allows net flux of calcium into bone. One of the preoperative clues to the likelihood of this syndrome is increased bone turnover as signaled by increased bone formation markers such as alkaline phosphatase (Table 4.6). The pathogenesis, diagnosis and management of hypocalcemia and the "hungry bones" syndrome are discussed in detail in Chapter 4, *Primary Hyperparathyroidism.*

Patients with activating CSR mutation have relative hypoparathyroidism in that their PTH levels are inappropriately low for their low serum calcium levels. The low PTH contributes to the nephrocalcinosis commonly seen in these patients. Patients with familial hypoparathyroidism have also been described with mutations of the noncoding region of the PTH gene.

■ Treatment

Calcium and vitamin D are used to treat hypoparathyroidism. Vitamin D is used to establish a baseline calcium and calcium is added (or subtracted) for acute changes in calcium. Calcitriol is the theoretically preferred preparation of vitamin D since it is the form of the hormone whose production is inhibited by the decreased PTH, and it is practically preferred because any toxic actions can be rapidly reversed, in contrast to the other forms of vitamin D (Table 5.4). Dihydrotachysterol (DHT) is an analog that is only 3-hydroxylated. Its properties approximate those of calcifediol. It undergoes 25-hydroxylation when taken and is considered to assume a pseudo 1α configuration in its receptor interaction. Although not currently approved by the Food and Drug Administration (FDA), 1α-hydroxylated forms (alphacalcidiol) of vitamin D_2 (Lunar) and D_3 (Leo) are used in other countries. Their characteristics are intermediate to those of the dihydroxylated analogs.

TABLE 5.4 — VITAMIN D PREPARATIONS

Generic/*Trade* Name	**Manufacturer**	**Daily Dose**	**Reaction Half-Time** ($T\frac{1}{2}$)	**Time for Normocalcemia**	**Duration of Action**
Ergocalciferol (vitamin D_2)/*Many*	Many	400 units	30 days	4 to 8 weeks	2 to 6 months
Cholecalciferol (vitamin D_3)/*Many*	Many	400 units	30 days	4 to 8 weeks	2 to 6 months
Calcifediol (25-OHD_3)/*Calderol*	Organon	1 to 5 µg	15 days	2 to 5 weeks	1 to 2 months
Calcitriol (1,25-DHCC)/*Rocaltrol*	Roche	0.25 to 0.5 µg	6 hours	2 to 5 days	1 to 2 days

5

Other analogs of vitamin D with both side-chain and ring modifications are used for their immunomodulatory properties, notably for psoriasis.

Several preparations of calcium are available. Since they all can provide the requisite amount of elemental calcium, clinical tolerance and expense are deciding criteria for their use (Table 5.5). A dose of approximately 1 g/d (preferably as divided doses for smoother control of calcium) should be adjusted to the lowest dose that keeps the patient asymptomatic.

Treatment with calcium and vitamin D may be needed only while injured parathyroid glands recover, or may be required chronically when all glands are removed and/or permanently injured. In the former cases, a traumatized parathyroid remnant recovers and resumes secretion of PTH. In some patients, hypocalcemia may not become evident until years after the procedure, presumably due to progression of surgical injury to the parathyroid glands. Hypocalcemia fol-

TABLE 5.5 — CALCIUM PREPARATIONS

	Grams Necessary to Provide 1 g of Elemental Calcium
Carbonate	2.5
Chloride	3.7
Acetate	4.0
Citrate	5.0
Glycerophosphate	5.7
Levulinate	7.7
Lactate	7.7
Orthophosphate	9.0
Gluconate	11.1
Glubionate	15.2

lowing neck surgery is least likely to ensue when only one of the parathyroid glands is injured or removed, as in the case of removal of a single adenoma; and most likely to ensue when all of the parathyroid glands are injured or removed (as in radical surgery), removal of three and one half hyperplastic glands for parathyroid hyperplasia, and aggressive exploration of all parathyroid glands during a difficult surgery.

Hypomagnesemia should always be considered as a potential contributory cause of hypocalcemia since PTH acts to maintain serum magnesium. A low serum magnesium may reveal the magnesium deficit, but the serum magnesium may be normal or low normal, since serum magnesium does not accurately reflect the stores of this primarily intracellular ion. Therefore, a therapeutic trial of magnesium, usually parenteral, may be needed to rule out magnesium deficiency.

Despite attaining normocalcemia, many patients with hypoparathyroidism remain symptomatic, especially females during menses. This points out that vitamin D and calcium treatment are not optimal agents for treatment. PTH would be the treatment of choice for most patients with hypoparathyroidism. The hormone induces a sense of well being in deficient patients and does not present the same risk of hypercalciuria as do calcium and vitamin D. However, PTH is not yet available as a treatment for hypoparathyroidism.

The objective of chronic therapy is to keep the patient free of symptoms and to maintain a serum calcium of approximately 8.5 to 9.2 mg/dL. With lower serum calcium levels, the patient may not only experience symptoms but may be predisposed over time to cataracts. With serum calcium concentrations in the upper normal range, there may be marked hypercalciuria, which occurs because the hypocalciuric effect of PTH has been lost. This may predispose to nephrolithiasis, nephrocalcinosis, and chronic renal insufficiency, so it becomes important to adjust medication doses for optimal effect. This is easier to accomplish by keeping the dose of vitamin D constant and fine tuning with calcium.

Both short-acting preparations of vitamin D, such as calcitriol and the synthetic analog dihydrotachysterol (DHT), and long-acting preparations such as vitamin D_3 (cholecalciferol) and vitamin D_2 (ergocalciferol) are available (Table 5.4). By far, the most inexpensive regimens are those that use cholecalciferol or ergocalciferol. In addition to economy, they have the advantage of rather easy maintenance in most patients. The disadvantage is that these preparations can slowly accumulate and produce delayed and prolonged vitamin D intoxification that can take a long time to reverse. Short-acting preparations are less prone to some of these effects but may require more frequent titration and are much more expensive than plain vitamin D. Short-acting preparations of vitamin D should be used in divided doses. Caution must be exercised in the introduction of other drugs that influence calcium metabolism. For example, diuretics can be both hypercalciuric (eg, furosemide, ethacrynic acid) and hypocalciuric (eg, thiazides).

Idiopathic Hypoparathyroidism

Idiopathic hypoparathyroidism can occur by itself or in association with polyglandular deficiency (Table 5.6), most commonly with gonadal, thyroid, and adrenal insufficiency. The parathyroid glands can be absent, remnant, or compromised by an immune infiltrate. Circulating parathyroid gland antibodies may be present, but their pathogenetic role is not established.

TABLE 5.6 — POLYENDOCRINE DEFICIENCY DISORDERS ASSOCIATED WITH HYPOPARATHYROIDISM

- Addison's disease
- Hypothyroidism (thyroiditis)
- Hypogonadism
- Hypergonadotropic hypogonadism (gonaditis)
- Vitiligo
- Alopecia
- Hypophysitis
- Ectodermal dystrophy
- Malabsorption (steatorrhea)
- Asplenia
- Chronic active hepatitis
- Cholelithiasis
- Pernicious anemia
- Diabetes (type 1)
- Mucocutaneous candidiasis
- Graves' disease
- Keratoconjunctivitis

Pseudohypoparathyroidism

Pseudohypoparathyroidism (PHP) is a genetic disorder characterized by target-organ unresponsiveness to PTH. PHP mimics the hormone-deficient forms of hypoparathyroidism, with hypocalcemia and hyperphosphatemia, but the PTH level is elevated rather than low or absent. PHP is classified as type 1 or type 2. There are several distinct forms of PHP1 (Table 5.7). Pseudohypoparathyroidism type 1B (PHP1B) is due to end-organ resistance to PTH caused by a heterozygous mutation in the GNAS1 gene. Pseudohypoparathyroidism type 1A (PHP1A) has in addition a characteristic somatic phenotype known as Albright's hereditary osteodystrophy (AHO), which consists of short stature, a round face, short neck, brachytactyly and subcutaneous calcifications. The

TABLE 5.7 — FEATURES OF PSEUDOHYPOPARATHYROIDISM TYPE 1

	PHP1A	PHP1B	PHP1C
Response to PTH	Defective	Defective	Normal
AHO present	Yes	No	Yes
G_s Mutation	Yes	No	No
Other hormone resistance	Yes	No	Yes

Abbreviations: PHP, pseudohypoparathyroidism; PTH, parathyroid hormone; AHO, Albright's hereditary osteodystrophy; G_s, stimulatory guanine nucleotide-binding protein.

shortening of the metacarpals (and metatarsals) causes the affected digits to have a dimple instead of a knuckle when a fist is made.

Primary hypothyroidism occurs commonly in patients with PHP, as does a gonadotropin-based abnormality of reproductive function — oligomenorrhea in females and infertility in males. Some individuals in families with PHP have AHO without any abnormalities of calcium metabolism. This disorder is called pseudopseudohypoparathyroidism.

5

Pseudohypoparathyroidism type 1A is due to an unresponsiveness of PTH to its receptor. This autosomal dominant genetic abnormality of the receptor complex is caused by loss of one functional allele of the gene encoding the stimulatory (s) G protein subunit G_s. These patients have a 50% deficiency of the heterotrimer G_s that couples the PTH receptor to adenylyl cyclase and a markedly blunted response of urinary cyclic adenosine monophosphate (cAMP) to the administration of PTH. This resistance is the basis of the Ellsworth-Howard test, where the patient is challenged with PTH. Normally, PTH administration increases urinary cAMP to levels greater than 300 nmol/L glomerular filtrate, but in PHP this response is absent or blunted (PTH is not readily available for this use). The response of urinary phosphate to PTH is also blunted, but this is a much less reliable test because of its nonspecificity.

Since G_s also couples many other receptors to adenyly cyclase, there can be associated primary hypothyroidism and primary hypogonadism, likely due respectively to resistance to thyroid-stimulating hormone (TSH), luteinizing hormone (LH), and follicle-stimulating hormone (FSH). However, response to other peptide hormones is apparently normal. Since G_s is also deficient in patients with pseudopseudohypoparathyroidism who have AHO, G_s mutation only sometimes produces resistance to PTH, suggesting other molecular mechanisms for the PTH resistance.

In PHP1B, characterized by resistance to PTH but no somatic phenotype, G_s is normal. Although also characterized by a blunted response to PTH, the molecular basis and inheritance pattern of this disorder is unknown. PTH type 1C has features of both, and G_s function is normal.

Pseudohypoparathyroidism type 2 is the designation for the disorder when characterized by a normal cAMP response to PTH but a decreased phosphaturic response and the absence of phenotypic or other hormonal deficiencies. Some patients exhibit vitamin D deficiency.

Hypoparathyroidism and Polyendocrine Failure

Hypoparathyroidism can occur in an autoimmune setting (Table 5.6) associated with parathyroid antibodies. The most commonly associated disorders are Addison's disease and candidiasis. In patients with Addison's disease and hypoparathyroidism, male and female patients are usually affected in childhood. Mucocutaneous candidiasis frequently precedes both endocrinopathies, and most patients with this infection and an associated endocrinopathy have hypoparathyroidism. Hypoparathyroidism develops in the first decade in the majority of patients. Addison's disease follows in about 2 years, and in most patients it occurs within 9 years of the onset of the syndrome. These patients can be affected by other endocrinopathies, including thyroid disease, diabetes mellitus, pernicious anemia, and ovarian failure (Table 5.6).

The polyendocrine disorders are divided into type 1 and type 2. Although there is some crossover, hypoparathyroidism is more common in type 1, which includes Addison's disease, mucocutaneous candidiasis, and dental hypoplasia. Type 2 is sometimes referred to as Schmidt's syndrome. While patients with idiopathic hypoparathyroidism frequently have circulating autoantibodies to parathyroid tissue, patients with Addison's disease and hypoparathyroidism also have an increased frequency of thyroid and parietal cell auto antibodies. Patients with isolated hypoparathyroidism, like those with Addison's disease, have an early onset of disease and also associated candidiasis.

Acute Hypocalcemic Disorders

Failure of normal calcium homeostasis can also cause hypocalcemia acutely. Hyperphosphatemia that results from phosphate administration, rhabdomyolysis, or tumor lysis, can produce severe hypocalcemia, especially in renal insufficiency. In acute pancreatitis, sequestration of calcium by saponification with fatty acids cause hypocalcemia. Rapid or excessive skeletal mineralization can cause hypocalcemia as in the "hungry bone" syndrome and in osteoblastic metastases (Table 5.2).

Treatment of Acute Hypocalcemia

The treatment of chronic hypocalcemia has been discussed above. Hypocalcemia can also be an endocrine emergency requiring rapid intravenous intervention. Care must be taken as intravenous calcium can be irritating to the veins, so the line should be fully open. Patients with severe hypocalcemia or with neurological manifestations or stridor should receive intravenous calcium at the approximate rate of 1 to 2 mg/kg over several minutes. This can be given with several preparations, including:

- Calcium chloride (272 mg calcium/10 mL)
- Calcium gluconate (90 mg calcium/10 mL).

Approximately 100 mg to 200 mg of elemental calcium can be given over several minutes. It should be noted that calcium and bicarbonate are not compatible solutions for admixture.

Oral calcium and a rapidly acting preparation of vitamin D should be started as soon as practical. If necessary, calcium can be given intravenously in doses of several hundred mg/24 hours until oral therapy has taken effect. Doses notwithstanding, treatment must be assessed with frequent serum calciums. Patients taking digitalis are predisposed to cardiotoxicity by infusion of calcium, therefore, an electrocardiogram (EKG) should be used for cardiac monitoring. Table 5.8 summarizes treatment protocol for hypocalcemia. More details about the acute treatment of post-PTX hypocalcemia are discussed in Chapter 4, *Primary Hyperparathyroidism.*

Magnesium deficiency should also be treated when present since it can attenuate the calcemic effect of the treatment by calcium and vitamin D (Table 5.9). As with calcium treatment, serum magnesium

TABLE 5.8 — TREATMENT OF HYPOCALCEMIA

Emergency	• 20 to 30 mL of 10% calcium (9 mg/mL) gluconate (100 mL D_5W) in 10 to 15 minutes IV
Intermediate	• Titrate serum calcium toward normal every 6 to 12 hours with 500 mg calcium in 1000 mL D_5W
Maintenance	• Calcium (many preparations, Table 5.5) • Vitamin D and magnesium as indicated • Magnesium: diarrheagenic

Abbreviations: D_5W, dextrose 5% in water (solution); IV, intravenous.

TABLE 5.9 — TREATMENT OF HYPOMAGNESEMIA

- Oral for mild, chronic magnesium deficiency:
  - Many preparations available, including:
    - Magnesium oxide (Mag-Ox 400, Blaine), 240 mg of elemental magnesium
    - Magnesium carbonate (Mag-Carb, R&D), 70 mg elemental magnesium
  - 200 to 300 mg/day
- Parenteral for severe hypomagnesemia:
  - Both 10% and 50% solutions of magnesium sulfate ($MgSO_4$, SoloPak and Astra) are available
  - 2 to 4 mL IV (or IM) of a 50% (48 mg) solution given over 10 to 15 minutes followed by similar amounts given daily
  - Several days of treatment are usually required to restore magnesium status

(and calcium) should be monitored frequently. However, since it is primarily an intracellular ion, serum measurements may not be reliable for monitoring therapy.

Summary

Renal failure and hypoparathyroidism are the most common causes of hypocalcemia, with hypoalbuminemia being a common artifactual cause. The most common clinical manifestations of hypocalcemia are due to increased neuromuscular irritability, either latent or actual, caused by the low calcium. Chvostek's and Trousseau's signs are classical, and tetany and convulsions are the extreme manifestations of hypocalcemia.

The etiologies of hypocalcemia can be generally classified as nonparathyroid and parathyroid. Thus the measurement of serum PTH and phosphorus are key procedures in differential diagnosis. In hypoparathyroidism, PTH is low or absent and serum phosphorus is increased; the opposite occurs in the nonparathyroid causes because of compensatory secondary hyperparathyroidism. An exception to this rule is pseudohypoparathyroidism, the genetic disease of end-organ resistance to PTH characterized by the biochemical features of hypoparathyroidism, a characteristic somatotype, but with an increased serum PTH.

Hypoparathyroidism can be part of an autoimmune endocrine deficiency syndrome that variably involves, among other tissues, the adrenal, thyroid, and pancreas. Magnesium deficiency is a relatively common cause of hypocalcemia in the hospitalized patient. Calcium and vitamin D supplements are used to treat most forms of hypocalcemia. The serum calcium should be kept at the low normal range in order to avoid hypercalciuria, which can occur without hypercalcemia, especially in the absence of PTH. Magnesium replacement may also be necessary in refractory hypocalcemia.

SUGGESTED READING

Arnold A, Horst SA, Gardella RJ, Baba H, Levine MA, Kronenberg HM. Mutation of the signal peptide-encoding region of the preproparathyroid hormone gene in familial isolated hypoparathyroidism. *J Clin Invest.* 1998;86:1084-1987.

Betterle C, Greggio NA, Volpato M. Clinical review 93: autoimmune polyglandular syndrome type I. *J Clin Endocrinol Metab.* 1998;83:1049-1055.

Hatley TK, Wheeler JG, Burks AW, Jones SM. Clinical spectrum of disease in patients with DiGeorge anomaly. *J Investig Med.* 1998;46:28A.

Iiri T, Farfel Z, Bourne HR. G-protein diseases furnish a model for the turn-on switch. *Nature.* 1998;394:35-38.

Ismail EA, al-Shammari N, Nadi H. Transient hypocalcemia with elevated serum parathormone in an infant of a hyperparathyroid mother. *Acta Paediatr Jpn*. 1998;40:290-292.

Kitanaka S, Takeyama K, Murayama A, et al. Inactivating mutations in the 25-hydroxyvitamin D3 1α -hydroxylase gene in patients with pseudovitamin D-deficiency rickets. *N Engl J Med*. 1998;338:653-661.

Marx SJ, Agarwal SK, Kester MB, et al. Germline and somatic mutation of the gene for multiple endocrine somatic mutation of the gene for multiple endocrine neoplasia type 1 (MEN1). *J Intern Med*. 1998;243:447-453.

McPherson ML, Prince SR, Atamer ER, Maxwell DB, Ross-Clunis H, Estep HL. Theophylline-induced hypercalcemia. *Ann Intern Med*. 1986;105:52-54.

Neufeld M, Maclaren NK, Blizzard RM. Two types of autoimmune Addison's disease associated with polyglandular autoimmune (PGA) syndromes. *Medicine*. 1981;60:355-362.

Parkinson DB, Thakker RV. A donor splice site mutation in the paratyroid hormone gene is associated with autosomal recessive hypoparathyroidism. *Nat Genet*. 1992;1:149-152.

Prendiville S, Burman KD, Wartofsky L, Ringel MD, Sessions RB. Evaluation and treatment of post-thyroidectomy hypocalcemia. *The Endocrinologist*. 1998;8:34-40.

Taylor HC, Ali MY. Transient ionized hypocalcemia and secondary hyperparathyroidism accompanying acute adrenal insufficiency. *Endocr Pract*. 1998;4:159-164.

Watanabe T, Bai M, Lane CR, et al. Familial hypoparathyroidism: identification of a novel gain of function mutation in transmembrane domain 5 of the calcium-sensing receptor. *J Clin Endocrinol Metab*. 1998;83:2497-2502.

Winer KK, Yanovski JA, Cutler GB Jr. Synthetic hyman parathyroid hormone 1-34 vs calcitriol and calcium in the treatment of hypoparathyroidism. *JAMA*. 1996;276:631-636.

6 Multiple Endocrine Neoplasia

Introduction

There are three multiple endocrine neoplasia (MEN) syndromes:

- MEN type 1 (also called Wermer's syndrome)
- MEN type 2A (also called Sipple's syndrome)
- MEN type 2B (sometimes called MEN type 3).

Their features are summarized in Table 6.1. In general, the clinical and biochemical characteristics of component tumors of MEN syndromes are similar to the characteristics of each tumor when it occurs by itself, but management is more complex. The glands most commonly involved in MEN are:

- Parathyroid
- Pituitary
- Pancreas
- Thyroid
- Adrenal.

The cell types involved in these tumors have in common an embryologic precursor in the neuroectoderm and/or the presence of the metabolic pathway for amine precursor uptake and decarboxylation (APUD). Oncogenic and mutational factors also influence expression of these tumors. The MEN syndromes are usually transmitted as autosomal dominant abnormalities, but there is considerable variability in penetrance and specific tumor incidence. Identification of the responsible genes and the availability of genetic testing have begun to revolutionize the diagnosis and man-

TABLE 6.1 — COMPONENTS OF THE MULTIPLE ENDOCRINE NEOPLASIA
MEN Type 1 • Parathyroid tumors 80% • Pituitary tumors 75% • Pancreatic tumors 65%
MEN Type 2A • Medullary thyroid carcinoma > 90% • Hyperparathyroidism > 50% • Pheochromocytoma < 20%
MEN Type 2B • Medullary thyroid carcinoma > 90% • Pheochromocytoma > 50% • Multiple mucosal neuroma (MMN), ganglioneuromatosis, and marfanoid habitus > 50%
Abbreviations: MEN, multiple endocrine neoplasia.

agement of patients with inherited endocrine tumors. Biochemical tests can be used to demonstrate the presence of the tumor. Imaging tests are generally less sensitive than biochemical tests and should be used when the biochemical tests are positive.

Clinical Features and Pathogenesis

■ Multiple Endocrine Neoplasia Type 1

The endocrine tumors in MEN type 1 and their approximate incidence is summarized in Table 6.2. The actual pattern may vary among kindreds. This disorder has been mapped to a gene on chromosome 11 that encodes a nuclear protein named menin, but the exact nature of the genetic defect has not yet been elucidated. Unlike MEN type 2, genetic diagnosis and pathogenesis is in its early stages in MEN type 1. So genetic screening for the many genetic mutations of this syndrome can only be conducted in a research set-

TABLE 6.2 — COMPONENTS OF MULTIPLE ENDOCRINE NEOPLASIA TYPE 1 AND APPROXIMATE FREQUENCY OF OCCURRENCE
Hyperparathyroidism (80%)
Pancreatic Tumors (75%) • Gastrinomas – Benign 20% – Malignant 30% • Insulinomas – Benign 20% – Malignant 5% • Nonfunctional tumors – Benign < 5% – Malignant < 5%
Pituitary Tumors (65%) • Chromophobe or nonfunctional adenomas – Benign 40% – Malignant < 5% • Eosinophilic tumors or acromegaly (benign) 15% • Cushing's disease, basophilic 5% • Mixed and other types (benign) < 5% • Prolactin-secreting tumors > 5%
Other Tumors (20%) • Carcinoid and bronchial adenomas < 5% • Lipomas and liposarcomas 5% • Adrenocortical adenomas 10% • Thymic tumors – rare

ting. Clinical screening is thus done by testing for the specific organ disease, such as measuring calcium and PTH for hyperparathyroidism; pituitary tests, commonly prolactin, for a pituitary tumor; and gastrin for a gastrinoma in the pancreas or elsewhere. It is notable that serum chromogranin A measurement can be surrogate single test for many of the endocrine tumors of both MEN.

The penetrance of MEN type 1 increases rapidly with age, so that by age 20 it is greater than 50%. Consequently, screening should begin by age 15 and, as indicated, conducted annually.

Primary Hyperparathyroidism

Primary hyperparathyroidism (PH) is the most common endocrine neoplasia of MEN type 1. The pathology of the parathyroid glands varies from adenomatous to hyperplastic, and all parathyroid glands seem to be involved in the neoplastic process, even ectopic glands. Thus aggressive surgical therapy directed toward all of the abnormal glands may be necessary for effective treatment.

Pancreatic Tumors

The types of gastrin-producing tumors and their approximate incidence are summarized in Table 6.2. The neoplasia can occur with the inicated approximate incidence in the:

- Pancreas (60%)
- Duodenum (15%)
- Other gastrointestinal (GI) sites (20%).

Hyperplasia occurs in 6% of patients. The tumors can be malignant (60%) and metastatic (80%) or benign (35%) but multiple (50%). Only 10% are readily resectable at diagnosis.

Measurement by radioimmunoassay of the hormone products of these tumors in serum is an important diagnostic procedure. In addition, their secretory products can explain some of the associated findings. Examples are:

- Peptic ulcer diathesis and gastrin
- Hypoglycemia and insulin
- Secretory diarrhea and vasoactive intestinal polypeptide (VIP) and perhaps pancreatic polypeptide (PP).

Glucagonomas can be associated with bullous dermatitis, but the pathogenesis of this lesion is obscure. The production of excess gastrin leading to peptic ulceration by a pancreatic islet cell tumor is called Zollinger-Ellison (ZE) syndrome. The ZE syndrome more commonly occurs by itself, but about 25% of cases are part of MEN type 1.

Pituitary Tumors

The types of pituitary tumors in MEN type 1 and their approximate incidence is summarized in Table 6.2. Chromophobe adenomas are usually nonfunctioning and benign by histologic criteria, but can cause endocrine abnormalities by mass effect on adjacent endocrine cells. Serum chromogranin A and pituitary glycoprotein α subunit measurements can be increased in nonfunctioning pituitary tumors. Tumors producing growth hormone (GH), adrenocorticotropic hormone (ACTH), prolactin (PRL), and somatostatin (SS) have also been described in this syndrome, and their manifestations are classically related to their location and hormonal products.

6

Other Tumors

The tumors indicated in Table 6.2 can be found in patients with MEN type 1, but their link to this disorder is tenuous.

■ Multiple Endocrine Neoplasia Types 2A and 2B

Medullary Thyroid Carcinoma

The tumors of MEN type 2 are summarized in Table 6.3. Medullary thyroid carcinoma (MTC), a tumor of the calcitonin (CT)-producing C cells of the thyroid gland, occurs in both MEN types 2A and 2B (Table 6.1). MTC is usually a firm, rounded tumor located in the middle or upper lobes of the thyroid

TABLE 6.3 — COMPONENTS OF MULTIPLE ENDOCRINE NEOPLASIA TYPE 2 AND APPROXIMATE FREQUENCY OF OCCURRENCE
MEN Type 2A • Medullary thyroid carcinoma (97%) • Hyperparathyroidism (50%) • Pheochromocytoma (30%)
MEN Type 2B • Multiple mucosal neuromas (99%) • Medullary thyroid carcinoma (90%) • Marfanoid habitus (65%) • Pheochromocytoma (45%)

gland. Because of its variable features, the histopathological diagnosis of MTC can be conclusively established only by immunohistological demonstration of CT in the abnormal C cells. Fine-needle aspiration (FNA) is usually not adequate for this purpose. The frank malignancy of MTC can be preceded by a progressive hyperplasia of C cells, referred to as C-cell hyperplasia, at least in familial cases. This precursor can develop in early childhood or as late as the second decade. There may be corresponding changes in the parathyroid and adrenal glands, and several instances of C-cell adenoma have also been reported.

Since the natural history of MTC can vary greatly, decisions regarding therapy are complex. Tumor behavior stands between the aggressive behavior of anaplastic thyroid carcinoma and the indolent behavior of papillary and follicular thyroid carcinoma. The most common sign of MTC is a thyroid nodule, and the most common symptom is diarrhea.

Calcitonin and Related Peptides and Proteins

The neoplastic C cells of the thyroid gland produce abnormally high amounts of CT. The basal concentrations of the hormone in serum are diagnostically elevated in many patients with MTC. Serum CT can also be increased in a few rare neuroendocrine tumors and perhaps in some acute illnesses. However, basal levels of CT are indistinguishable from normal in an increasing percentage of patients with this tumor as the diagnosis is being recognized earlier. Thus, provocative tests with calcium and/or pentagastrin have been developed for the diagnosis of MTC and its histologic antecedents. Serial measurements of serum CT can be used to monitor therapy. Other products of the CT gene, notably the 37-amino-acid peptide termed calcitonin gene-related peptide (CGRP), can also be increased (Figure 2.1). In addition to CT, MTC produces other substances, nonpeptides as well as peptides (Table 6.4).

Pheochromocytoma

Pheochromocytoma is a component of MEN types 2A and 2B. Bilateral and multifocal pheochromocytomas occur with an incidence of over 70% in these clinical settings. By contrast, bilateral incidence is usually less than 10% for sporadic pheochromocytomas and only 20% to 50% in familial pheochromocytomas. The MTC may precede the pheochromocytoma by as many as 2 decades, and another pheochromocytoma may occur after removal of the first.

Adrenal medullary hyperplasia is a precursor of the pheochromocytomas seen with MTC. The usual diagnostic tests for pheochromocytoma should be pursued vigorously, because the manifestations of this tumor may be subtle. Thus, the only biochemical abnormality may be an abnormal ratio of urinary (or serum) epinephrine to norepinephrine.

Primary Hyperparathyroidism

Primary hyperparathyroidism is more common in MEN type 2A than in MEN type 2B (Table 6.1). Hy-

TABLE 6.4 — PRODUCTS SECRETED BY MEDULLARY THYROID CARCINOMA
• Adrenocorticotropic hormone • Amyloid • Calcitonin • Calcitonin gene-related peptide • Carcinoembryonic antigen • Catecholamines • Chromogranin A • Corticotropin-releasing hormone • Dopa decarboxylase • β-Endorphin • Histaminase • Kallikrein and kinins • Melanin • Melanocyte-stimulating hormone • Nerve growth factor • Neuron-specific enolase • Neurotensin/gastrin-release peptide • Prolactin-releasing hormone • Prostaglandins • Serotonin • Somatostatin • Substance P • Vasoactive intestinal polypeptide

perplasia is more common than adenoma (see Chapter 4, *Primary Hyperparathyroidism*).

Mucosal Neuromas

The presence of neuromas with a centrofacial distribution is the most consistent component of MEN type 2B. The most common location of neuromas is the oral cavity (tongue, lips, buccal mucosa), but other sites may be involved. GI tract abnormalities occur in the multiple mucosal neuroma (MMN) syndrome, the most common being GI ganglioneuromatosis. The anatomic GI abnormalities can be associated with functional difficulties in swallowing, megacolon, diarrhea, and constipation.

Marfanoid Habitus

A marfanoid habitus is seen commonly in the MMN syndrome, but unlike true Marfan's syndrome, such patients do not have aortic arch abnormalities, ectopia lentis, homocystinuria, or mucopolysaccharide abnormalities.

Genetic Basis of MEN

The genetic basis for the MEN has become elucidated, especially for MEN type 2. For MEN type 1, the defect has been localized to a site on chromosome 11 and mutations in the gene termed menin. Menin mutations are scattered throughout the gene and not yet susceptible to clinically practicable diagnostic testing. No genotype-phenotype correlations have been established as have been for MEN type 2. MTC, which accounts for about 5% to 10% of all thyroid malignancies, is usually sporadic, with genetic abnormalities accounting for approximately 25% of tumors.

The pathogenesis of hereditary MTC has been elucidated by the identification of causative mutations in the RET protooncogene, which resides in chromosome 10 and encodes a tyrosine kinase receptor for a putative neurotropic growth factor. Two broad classes of mutation have been identified, involving respectively intracellular and extracellular domains of the tyrosine kinase receptor. Different mutations in RET can produce several distinct diseases. Inheritance of activating (gain-of-function) mutations is responsible for MEN type 2A and familial MTC; inheritance of a different set of activating mutations causes MEN type 2B.

The elucidation of the genetic basis of MEN has led to the discovery of families with the inherited disorder that had been mistakenly classified as sporadic. Thus, although activating germinal mutations of RET account for more than 90% of hereditary cases of MTC, up to 10% of apparently sporadic MTCs may actually be familial. In addition, somatic RET mutations are found in approximately 25% of sporadic MTCs, and these tumors appear to have a more aggressive course, especially those of codon 918. RET mutations of codons 768 and 609 are also found in familial isolated MTC (FMTC).

Mutations that inactivate (loss-of-function) the RET gene produce Hirschsprung's disease, a congenital absence of the enteric parasympathetic ganglia resulting in megacolon, and a skin disease involving the upper back descriptively named cutaneous lichen amyloidosis (CLA). Elucidation of genetic pathogenesis has led to an expansive reclassification of this group of disorders (Table 6.5).

TABLE 6.5 — MULTIPLE ENDOCRINE NEOPLASIA TYPE 2 VARIANTS
• MEN type 2A • MEN type 2A with Hirschsprung's disease • MEN type 2A with cutaneous lichen amyloidosis (CLA) • Familial medullary thyroid carcinoma (FMTC) • MEN type 2B

Genetic and Biochemical Diagnosis and Management

Both genetic testing for RET proto-oncogene mutations and biochemical testing for elevated serum CT have a role in clinically managing patients with MEN type 2 and their families. Genetic testing with its sensitivity and specificity is the standard for diagnosis. Approximately a dozen mutations of RET account for most cases; screening for them is practical and commercially available.

Calcitonin measurement can be a confirmatory diagnostic procedure for demonstrating the actual presence of the disease as opposed to its genetic probability. Some have recommended that all patients with thyroid nodules have serum CT measurements, estimating that approximately one half of 1% of all thyroid nodules represent MTC. Another still important and unique role for CT measurement is to monitor therapy of MTC.

Treatment of Multiple Endocrine Neoplasia Syndromes

In general, the individual endocrine components of MEN syndromes have the same clinical and biochemical manifestations as when they occur as individual tumors. However, the manner in which the individual presents to the physician is a unique aspect of MEN. If the patient is part of a MEN kindred, the physician must evaluate the patient for each potential component of the syndrome. Or the physician may search for the presence of MEN when a patient presents with one of the syndrome's endocrine tumors, although the yield may not be great. The diagnoses of one patient with MEN can lead to the identification of other involved relatives.

Surgery is the treatment of choice for the neoplasias in MEN. Although some neoplasias are potentially lethal, especially MTC and pheochromocytoma, all can be cured by early diagnosis followed by operation. Management of the individual components of MEN follows the standard procedures for each of the neoplasias.

Pheochromocytomas, which are commonly bilateral, especially in MEN type 2B should be treated first in order to obviate their cardiovascular threats. Surgery should be aggressive, because all glandular tissue may be involved. Thyroidectomy should be conducted as early as practical. Metastatic disease has been reported at age 5 in MEN type 2A and at age 2 in the more aggressive MEN type 2B. Since MEN syndromes are transmitted in an autosomal dominant pattern, family members should be evaluated. The duty of the physician to inform family members who are not the physician's patients of the presence of familial disease can present both legal and ethical issues.

Summary

Neoplasia of the parathyroid, pituitary, pancreas, thyroid, and adrenal can occur in inherited syndromes involving multiple endocrine glands. There are three such syndromes of multiple endocrine neoplasia (MEN):

- MEN type 1: parathyroid adenomas and hyperplasia, pituitary adenomas, usually nonfunctioning, and pancreatic tumors, most commonly benign insulinomas and gastrinomas
- MEN type 2A: medullary thyroid carcinomas, parathyroid adenomas and hyperplasia, and pheochromocytomas
- MEN type 2B: medullary thyroid carcinomas, pheochromocytomas, and multiple mucosal neuromas (MMN).

Variants have been described of MEN type 2. These syndromes are inherited in an autosomal dominant pattern. Mutation of the RET proto-oncogene, encoding a tyrosine kinase, is the genetic basis of MEN type 2. MEN type 1 has been recently localized to a gene on chromosome 11 that encodes a protein of uncertain function termed menin. The genetic mutations of these syndromes form the basis of genetic testing and family screening. Such testing can raise both ethical and legal issues. For MEN type 2, genetic testing can focus on the approximate dozen mutations that account for essentially all cases. For MEN type 1, genetic testing has not yet developed to general clinical application.

Measuring the endocrine products of the involved endocrine tissue in serum or urine can also be used in diagnosis and to monitor therapy. For MEN type 1, these measurements include parathyroid hormone (PTH) (and calcium), pituitary hormones (including prolactin), and gastrin; for MEN type 2, calcitonin (CT), PTH (and calcium), and catecholamines. Surgical treatment for each of the tumors is generally similar to such treatment for the individual tumors when they occur alone. However, a pheochromocytoma should be treated first because of its potential for cardiovascular complications. Furthermore, the surgical approach should address the frequent bilateral tumors that occur in these syndromes.

SUGGESTED READING

Bassett JHD, Forbes SA, Pannett AAJ, et al. Characterization of mutations in patients with multiple endocrine neoplasia type 1. *Am J Hum Genet*. 1998;62:232-244.

Chandrasekharappa SC, Guru SC, Manickam P, et al. Positional cloning of the gene for multiple endocrine neoplasia-type 1. *Science*. 1997;276:404-407.

Deftos LJ. Calcitonin and medullary thyroid carcinoma. In: Bennett JC, Plum F, eds. *Cecil Textbook of Medicine*. 20th ed. Philadelphia, Pa: WB Saunders Company; 1996:1332-1335.

Deftos LJ. Genomic torts: the law of the future—the duty of physicians to disclose the presence of a genetic disease to the relatives of their patients with the disease. *Univ of San Francisco Law Review*. 1997;32:105-137.

Deftos LJ, Bury AE, Habener JF, Singer FR, Potts JT Jr. Immunoassay for human calcitonin. II. Clinical studies. *Metabolism.* 1971;20:1129-1137.

Deftos LJ, Nolan JJ. Syndromes involving multiple encodrine glands. In: Greenspan FS, Strewler GJ, eds. *Basic and Clinical Endocrinology.* 5th ed. Norwalk, CT: Appleton & Lange; 1997:753-769.

Eng C, Clayton D, Schuffenecker I, et al. The relationship between specific RET proto-oncogene mutations and disease phenotype in multiple endocrine neoplasia type 2. International RET mutation consortium analysis. *JAMA.* 1996;276:1575-1579.

Gagel RF. Unresolved issues in the genesis and management of multiple endocrine neoplasia type 2. *Horm Metab Res.* 1997;29: 135-137.

Gagel RF, Cote GJ, Martins Bugalho MJ, et al. Clinical use of molecular information in the management of multiple endocrine neoplasia type 2A. *J Intern Med.* 1995;238:333-341.

Guru SC, Goldsmith PK, Burns AL, et al. Menin, the product of the MEN1 gene, is a nuclear protein. *Proc Natl Acad Sci U S A.* 1998;95:1630-1634.

Heshmati HM, Gharib H, van Heerden JA, Sizemore GW. Advances and controversies in the diagnosis and management of medullary thyroid carcinoma. *Am J Med.* 1997;103:60-69.

Kopp P. A long-sought needle in the haystack: the multiple endocrine neoplasia type 1 gene. *Eur J Endocrinol.* 1997;137:222-223.

Lallier M, St-Vil D, Giroux M, et al. Prophylactic thyroidectomy for medullary thyroid carcinoma in gene carriers of MEN2 syndrome. *J Pediatr Surg.* 1998;33:846-848.

Marx SJ, Agarwal SK, Kester MB, et al. Germline and somatic mutation of the gene for multiple endocrine neoplasia type 1 (MEN1). *J Intern Med.* 1998;243:447-453.

Marx S, Spiegel AM, Skarulis MC, Doppman JL, Collins FS, Liotta LA. Multiple endocrine neoplasia type 1: clinical and genetic topics. *Ann Intern Med.* 1998;129:484-494.

Mulligan LM, Kwok JB, Healey CS, et al. Germ-line mutations of the RET proto-oncogene in multiple endocrine neoplasia type 2A. *Nature.* 1993;363:458-460.

Niccoli P, Wion-Barbot N, Caron P, et al. Interest of routine measurement of serum calcitonin: study in a large series of thyroidectomized patients. *J Clin Endocrinol Metab.* 1997;82:338-341.

O'Connor DT, Deftos LJ. Secretion of chromogranin A by peptide-producing endocrine neoplasms. *N Engl J Med.* 1986;314: 1145-1151.

Pacini F, Fontanelli M, Fugazzola L. Routine measurement of serum calcitonin in nodular thyroid diseases allows the preoperative diagnosis of unsuspected sporadic medullary thyroid carcinoma. *J Clin Endocrinol Metab.* 1994;78:826-829.

Stjernholm MR, Freudenbourg JC, Mooney HS, Kinney FJ, Deftos LJ. Medullary carcinoma of the thyroid before age 2 years. *J Clin Endocrinol Metabol.* 1980;51:252-253.

Wells SA Jr, Skinner MA. Prophylactic thyroidectomy, based on direct genetic testing, in patients at risk for the multiple endocrine neoplasia type 2 syndromes. *Exp Clin Endocrinol Diabetes.* 1998;106:29-34.

Zedenius J, Larsson C, Bergholm U, et al. Mutations of codon 918 in the RET proto-oncogene correlate to poor prognosis in sporadic medullary thyroid carcinomas. *J Clin Endocrinol Metab.* 1995;80:3088-3090.

6

PART III

Major Diseases of Skeletal Metabolism

7

Osteomalacia and Rickets

Introduction

Osteomalacia, meaning "soft bones," is an abnormality of the mineralization of bone matrix in adults. When it occurs in children, the disorder is called rickets. Although the basic disorder is the same and both adults and children classically have associated hypocalcemia, hypophosphatemia, increased serum alkaline phosphatase, and increased serum parathyroid hormone (PTH), the clinical manifestations differ with age. In children, the mineralization of bone at the growth plate is impaired, producing wide cranial sutures, epiphysial dysplasia, widening and disorganization of the growth plate, and retardation of longitudinal bone growth. A variety of skeletal deformities are accentuated by the effects of gravity and muscle tension on the prone or supine infant or child. In adults, the effects of gravity and muscle tension are less and may result in only bowing abnormalities of the weight-bearing long bones. Intermediate stages are caused by the combined effects of abnormal mineralization, muscle pull, and gravity.

Pathology

The impaired mineralization of bone matrix, or osteoid, that is the hallmark of osteomalacia (and rickets) leads to the development of large surface areas of unmineralized matrix. Although specialized bone biopsy techniques may be needed to confirm the diagnosis, it can often be made on clinical grounds. Tetracycline, which concentrates at mineralizing sites, can be administered in conjunction with bone biopsy in order to label the mineralizing front. Two tetracycline labels can be administered 10 to 14 days apart in order to "quantify" mineralization. An undecalcified bone section will show thickened and increased osteoid seams between the two labels, reduced tetracycline labeling, and a reduction in the mineralization rate as measured by the double tetracycline labeling.

Etiology and Pathogenesis

Osteomalacia (and rickets) can be due to one or a combination of abnormalities of calcium, vitamin D, phosphate, or alkaline phosphatase (Table 7.1). The most common pathogenesis is a disorder of vitamin D availability, activation, or action (Table 7.2). Vitamin D abnormalities cause osteomalacia by resulting in inadequate amounts of calcium and phosphorus to support the normal mineralization of matrix. Similarly, primary or secondary calcium and phosphate deficiency states cause osteomalacia when their reduced ambient concentration leads to an inadequate supply of these minerals for proper matrix mineralization. Disorders that produce phosphaturia also limit the availability of phosphate for mineralization of bone and thus can cause osteomalacia and rickets. A rare cause is an inherited defect in the synthesis of alkaline phosphatase, an enzyme important in the mineralization process. Certain drugs in excess, including the bisphosphonates and sodium fluoride, also cause osteomalacia by their toxicity to the mineralizing osteoblast and the mineralizing front.

Symptoms and Signs

■ Clinical Manifestations

The clinical manifestations of osteomalacia or rickets depend on the severity and duration of the dis-

TABLE 7.1 — CAUSES OF OSTEOMALACIA AND RICKETS
Vitamin D and Calcium Abnormalities • Vitamin D and calcium deficiency (Table 7.2) – Dietary – Inadequate sunlight • Malabsorption of vitamin D and calcium • Abnormal vitamin D metabolism – Renal failure – Liver disease – Drugs • Hereditary vitamin D-vitamin D receptor (VDR) abnormalities – Vitamin D-dependent rickets type I (VDDR-I): defective 1α-hydroxylase – Vitamin D-dependent rickets type II (VDDR-II): mutations of the VDR • Deficient hormone binding to receptor • Deficient nuclear uptake of hormone-receptor complex • Deficient binding of vitamin D-VDR complex to regulatory DNA sequences
Hypophosphatemic Disorders • X-Linked hypophosphatemia • Autosomal recessive hypophosphatemia • Fanconi's syndrome • Renal tubular acidosis (type II)
Oncogenic Osteomalacia
Hypophosphatasia
Drug-Related Osteomalacia

order and the patient's age. Bone pain and bone tenderness are common in most patients. Patients with rickets have short stature because of impaired growth plate function in long bones and gravity-related deformities of bone that include a flattened pelvis, bowing of the long bones, varus or valgus deformities of

TABLE 7.2 — PATHOGENESIS OF VITAMIN D DISORDERS
Diet (↓ vitamin D) • Deficiency • Malabsorption • Gastrointestinal disease
Skin: Seasonal and Geographic Variations in Sunlight
Liver: Severe Enough Disease to Interfere With Vitamin D Metabolism (↓ 25-hydroxyvitamin D) • 25-hydroxylase defect • Anticonvulsant therapy
Kidney (↓ 1,25-dihydroxyvitamin D) • Chronic renal failure • 1α-hydroxylase defect (vitamin D-dependent rickets type 1 [VDDR-I])
Target Cell: Vitamin D Receptor Mutations (vitamin D-dependent rickets type II [VDDR-II])
End-Organ Resistant Rickets • Renal failure • Phenytoin

the knee, platybasia, and craniotabes. Also prominent are frontal bossing of the cranium, bulging of the costochondral junctions ("rachitic rosary"), and lateral indentation of the chest wall at the diaphragmatic insertion (Harrison's groove). In adults with osteomalacia, skeletal deformities are relatively mild, but bowing of the long bones of the legs is not uncommon. In vitamin D-deficient and hypophosphatemic osteomalacia, proximal muscle weakness can also be a prominent feature, producing a characteristic gait.

■ Laboratory Findings

The laboratory features of osteomalacia depend upon the etiology. In the classical syndromes, vitamin D levels are low as is urinary calcium. Patients have hypocalcemia, hypophosphatemia, and an increased serum PTH and alkaline phosphatase. The hypocalcemia and hypophosphatemia are due to the vitamin D abnormality, and the latter is accentuated by the phosphaturic effect that results from the secondary hyperparathyroidism. The increased alkaline phosphatase reflects the osteoblasts' abortive attempt to form new bone.

These classical features may vary in certain circumstances. Alkaline phosphatase levels are decreased in hypophosphatasia. The tendency toward hypocalcemia may be compensated for by the increased PTH early in the course of the disease. When there is impaired activation or biological action of vitamin D, there may be a compensatory increase in serum 1,25-dihydroxyvitamin D (1,25-D) levels. In vitamin D deficiency, the 25-hydroxyvitamin D (25-D) level is low, usually below 15 ng/dL.

Many of the other laboratory abnormalities can be related to the development of secondary hyperparathyroidism due to the hypocalcemia. The serum alkaline phosphatase and osteocalcin are usually high, reflecting the compensatory osteoblast activity that produces abundant but unmineralized osteoid. While the concentration of PTH is elevated, the serum 1,25-D is often normal, though a normal level may be homeostatically inappropriate to the hypophosphatemia, hypocalcemia, and secondary hyperparathyroidism.

■ X-ray Findings

The classic features of vitamin-D deficient osteomalacia and rickets are readily appreciated and can be diagnostically important. Epiphyseal dysplasia and delayed mineralization at the growth plate gives the x-ray appearance of widening of the epiphyses, fraying of the epiphysial ends, and widening and cupping of the metaphyses. The skeleton is generally under-mineralized and the cortices are thin. These latter findings can confuse osteomalacia with osteoporosis, a serious mistake since the treatment and effectiveness for the two conditions are so different.

An essentially diagnostic feature of osteomalacia is the presence of pseudofractures, also known as Looser's zones or Milkman's syndrome, zones of bone rarefaction at sites where pulsing arteries presumably erode bone at the point they cross it. The Looser's zones are often bilateral, as are the involved arteries, and most commonly occur at:

- Femoral neck
- Femoral shaft
- Pubic and ischial rami
- Clavicles, ribs, and scapulae
- The radius and ulna.

Signs of the secondary hyperparathyroidism, such as subperiosteal bone resorption, occur at the skull, phalanges and clavicles.

Etiologies of Osteomalacia and Rickets

■ Vitamin D Abnormalities

Vitamin D Deficiency

Vitamin D deficiency can occur when there are inadequate amounts of vitamin D in the diet or inadequate exposure to ultraviolet (UV) radiation. It usually develops only when both the diet and exposure to ultraviolet light are deficient. Only a short exposure to sunlight is needed to produce adequate amounts of vitamin D and milk is thus supplemented. However, both sunlight and diet can be inadequate in Northern climates and in elderly patients who are hospitalized or institutionalized. These circumstances notwithstanding, true deficiency of vitamin D is uncommon in the United States because of supplementation of milk with vitamin D.

Malabsorption of Vitamin D and Calcium

Malabsorption of vitamin D and consequently calcium may occur in any gastrointestinal (GI) abnormality where the absorption of fat soluble substances, like vitamin D, is impaired. These include:

- GI surgery
- Gall bladder disease
- Hepatic disease and impaired production of bile salts (hepatic osteodystrophy)
- Pancreatic insufficiency
- Diffuse disorders of the intestinal mucosa, such as celiac disease.

Abnormal Vitamin D Metabolism

The most common form of abnormal vitamin D metabolism that leads to osteomalacia is the impaired conversion of 25-D to 1,25-D that occurs in renal failure. Although theoretically possible, liver disease and resulting abnormalities in the conversion of vitamin D to its 25-hydroxy metabolite are rarely associated with significant osteomalacia, even in severe liver disease. Furthermore, osteomalacia is uncommon in patients treated with anticonvulsant drugs even though they enhance the oxidation of vitamin D in the liver and impair metabolic activation to 25-D. Impaired metabolism can also occur in vitamin D-dependent rickets type I (VDDR-I), a rare autosomal recessive trait due to deficiency of the renal 25-hydroxyvitamin D-1α-hydroxylase. Several gene mutations have been identified in this disorder.

Peripheral Resistance to Vitamin D

Vitamin D cannot mediate its biological effects when it does not correctly bind its receptor, the vitamin D receptor (VDR), and there are several diseases where this occurs. Vitamin D-dependent rickets type II (VDDR-II) is a rare autosomal recessive disorder caused by mutations in the vitamin D receptor that impair its binding and response to vitamin D. The resulting osteomalacia is characterized by a presumably compensatory increased in circulating levels of 1,25-D. In some patients, the disease is associated with epidermal abnormalities, including alopecia totalis. The many mutations of the VDR that have been identified in the syndrome can impair vitamin D binding to the receptor or prevent binding of the receptor to normally responsive DNA sequences in target genes. The healing of the bone disease in a few such patients by chronic intravenous infusions of calcium suggests that the role of vitamin D on bone formation is not direct but to provide adequate calcium (and phosphorus) for mineralization of matrix.

■ Hypophosphatemic Disorders

Like calcium, phosphate is important for matrix mineralization and its deficiency can result in osteomalacia and rickets. Hereditary and acquired disorders of renal phosphate wasting are thus among these causes. Familial X-linked hypophosphatemic rickets is the predominant of this type of cause of childhood rickets. The candidate gene encodes an endopeptidase whose role in the pathogenesis of the "phosphate diabetes" is not yet appreciated. Although the penetrance of the osteomalacia is variable in female carriers, affected males are severely rachitic. The serum phosphorus is very low because of renal phosphate wasting. While the concentration of 1,25-D in serum is often in the lower range of normal, it is inappropriately low for the hypophosphatemia. Serum calcium and PTH concentrations are usually normal. The disease responds to phosphate supplementation and to vitamin D, usually calcitriol, to prevent any secondary hyperparathyroidism.

Several other renal phosphate wasting conditions also produce hypophosphatemia osteomalacia. They include an autosomal recessive hypophosphatemic rickets, X-linked recessive hypophosphatemic rickets caused by mutations of the renal chloride channel, and X-linked recessive nephrolithiasis. Fanconi's syndrome, the generalized renal transport disorder, is also associated with rickets and osteomalacia. Abuse of phosphate-binding antacids can also cause hypophosphatemia and osteomalacia.

■ Oncogenic Osteomalacia

A rare cause of osteomalacia is termed oncogenic osteomalacia. It is a disorder of renal phosphate wasting and osteomalacia caused by mesenchymal tumors of bone and soft tissue that are often small, even unnoticeable, and benign or by prostate cancer. Such tumors secrete a "humoral factor" that causes the phosphaturia. Resection of the tumor often cures or dramatically improves the disease.

■ Hypophosphatasia

Alkaline phosphatase is involved as one of the molecular components of bone formation, so it is not unexpected that hypophosphatasia can cause rickets and osteomalacia. This rare deficiency of alkaline phosphatase may be inherited as an autosomal recessive or an autosomal dominant trait. The disorder is characterized by low alkaline phosphates activity, high serum and urinary levels of inorganic pyrophosphate and phosphoryl ethanolamine, and sometimes by hypercalcemia. The bone disease reflects the critical role of alkaline phosphatase in the mineralization process. Experimental treatments with infusions of alkaline phosphatase have been tried with varying degrees of success.

■ Drug-Related Osteomalacia

Several drugs can impair mineralization and cause osteomalacia. Sodium fluoride at doses of 75 mg/day can have this effect, while lower doses do not. The bisphosphonate edidronate disodium at doses of 10 to 20 mg/kg/d was associated with osteomalacia in early trials, but the prescribed dose of 5 mg seem relatively free of the effect. The newer bisphosphonates have a considerably reduced propensity to cause osteomalacia. Aluminum and tetracycline toxicity can also cause osteomalacia. Abuse of phosphate-binding antacids can deplete phosphorus stores and cause hypophosphatemia and osteomalacia. Phenytoin causes target organ resistance to 1,25-D in addition to increasing hepatic catabolism of 25-D.

Differential Diagnosis

The diagnosis of osteomalacia can be made by a consistent history and the appropriate pattern of serum calcium, phosphate, PTH, vitamin D metabolites and alkaline phosphatase. Bone biopsy can be diagnostic but is seldom needed. In the classical form, PTH and alkaline phosphatase are increased and calcium, phosphate, and vitamin D, especially 25-D, are decreased. It is important to emphasize that osteomalacia can sometimes be mistaken for osteoporosis. Such a mistake should be avoided by appreciating the fact that the dramatic biochemical abnormalities that occur in osteomalacia are not seen in osteoporosis, where routine biochemistries and serum levels of PTH and 1,25-D are usually normal.

Treatment of Osteomalacia

Treatment of osteomalacia is based on an understanding of the particular pathogenesis and the main four factors involved alone or in combination in virtually all cases. These are:

- Calcium
- Vitamin D metabolites
- Phosphate
- Alkaline phosphatase.

An appropriately adequate supply of calcium (Table 5.5) and phosphate (Table 12.7) should be provided for all patients. Simple vitamin D-deficient osteomalacia can be treated with replacement doses of vitamin D, 400 IU day (Table 5.4). If vitamin stores need to be repleted, higher doses may lead to more rapid healing. Administration of 50,000 IU/week for several weeks has been recommended to replete stores before maintenance is begun. Phosphate treatment is critical in phosphate-wasting disorders. Every attempt should be made to raise and maintain the serum phosphate to the normal range, although this may be difficult because of the diarrheal effects of most phosphate preparations. The administration of vitamin D and/or calcium may be necessary to prevent phosphate-induced hypocalcemia and to promote bone healing in these patients.

Calcitriol is the drug of choice in renal failure. It is available in oral forms (Table 5.4) that can be given on a daily basis and in parenteral forms that can be given intermittently, for example, at the time of dialysis. The injectable form of calcitriol (Calcijex) contains 1 or 2 micrograms of calcitriol in a 1 mL solution and is given in doses of 1 μg to 3 μg/week. Larger doses of vitamin D can be used to try to overcome the end-organ resistance of some of the VDR diseases, but hypercalcemia is always a risk with large doses of any vitamin D preparation. And finally, there have been some reports of infusions of alkaline phosphatase in patients with hypophosphatasia.

Summary

Osteomalacia is a group of skeletal diseases caused by the inadequate or abnormal mineralization of the organic matrix of bone, osteoid. The consequently "soft" skeleton cannot perform its mechanical function, and it exhibits:

- Gross structural abnormalities (eg, short stature)
- Specific defects due to the effects of gravity (eg, bowing)
- Abnormal muscle insertions (eg, Harrison's grooves)
- Signs of vascular pulsations (Looser's zones).

Bone pain and tenderness are common as are the clinical and biochemical manifestations of the hypocalcemia and resultant secondary hyperparathyroidism that commonly accompanies osteomalacia. Called rickets in children, the skeletal disease is due to a deficiency of one or more of four essentials for normal mineralization:

- Vitamin D metabolites
- Calcium
- Phosphorus
- Alkaline phosphatase.

Abnormalities in vitamin D and its activation and action are among the common causes of osteomalacia. Thus, any defect in the vitamin D metabolism pathway from intake to biological action on the receptor can cause osteomalacia and rickets. Similarly, any abnormality in the pathway of calcium and phosphorus absorption and renal excretion, such as malabsorption and hyperexcretion syndromes, can cause osteomalacia or rickets. Rare causes are inherited diseases characterized by:

- Inadequate retention of dietary phosphorus
- Defective alkaline phosphatase synthesis

- Abnormalities of the enzymes that metabolically activate vitamin D (VDDR-I)
- Abnormalities of the vitamin D receptor (VDDR-II).

The designation of vitamin D resistant rickets (VDRR) has been used for several of these disorders.

It is important to distinguish osteomalacia from osteoporosis because of the different clinical approaches to each. This can usually be accomplished by routine x-rays and biochemical evaluation. The former are distinctive for each disease, and where osteoporosis has normal biochemistries, osteomalacia is usually characterized by:

- Hypocalcemia
- Hypophosphatemia
- Increased alkaline phosphatase
- Decreased vitamin D metabolites.

7

Treatment is directed at the specific pathogenesis and involves adequate dietary calcium, phosphorus, and vitamin D.

SUGGESTED READING

Awumey EMK, Mita DA, Hollis BW, Kumar R, Bell NH. Vitamin D metabolism is altered in Asian Indians in the southern United States: a clinical research center study. *J Clin Endocrinol Metab.* 1998;83:169-173.

Beck L, Karaplis AC, Amizuka N, Hewson AS, Ozawa H, Tenenhouse HS. Targeted inactivation of Npt2 in mice leads to severe renal phosphate wasting, hypercalciuria and skeletal abnormalities. *Proc Natl Acad Sci U S A.* 1998;95:5372-5377.

Bouillon R. The many faces of rickets. *N Engl J Med.* 1998;338: 681-682. Editorial.

Gonzalez-Compta X, Manos-Pujol M, Foglia-Fernandez M, et al. Oncogenic osteomalacia: case report and review of head and neck associated tumours. *J Laryngol Otol*. 1998;112:389-392.

Hoshino H, Kushida K, Takahashi M, et al. Characteristics of biochemical markers in patients with metabolic bone disorders. *Endocr Res*. 1998;24:55-64.

Kitanaka S, Takeyama K, Murayama A, et al. Inactivating mutations in the 25-hydroxyvitamin D_3 1α-hydroxylase gene in patients with pseudovitamin D-deficiency rickets. *N Engl J Med*. 1998; 338:653-661.

Lenchik L, Sartoris DJ. Orthopedic aspects of metabolic bone disease. *Orthop Clin North Am*. 1998;29:103-134.

Marriott BM. Vitamin D supplementation: a word of caution. *Ann Intern Med*. 1997;127:231-233. Editorial.

Riggs BL. Vitamin D-receptor genotypes and bone density. *N Engl J Med*. 1997;337:125-126. Editorial.

Thomas MK, Lloyd-Jones DM, Thadhani RI, et al. Hypovitaminosis D in medical inpatients. *N Engl J Med*. 1998;338:777-783.

Zhu W, Malloy PJ, Delvin E, Chabot G, Feldman D. Hereditary 1,25-dihydroxyvitamin D-resistant rickets due to an opal mutation causing premature termination of the vitamin D receptor. *J Bone Miner Res*. 1998;13:259-264.

8 Paget's Disease of Bone (Osteitis Deformans)

Introduction

Paget's disease (PD) is a disorder of bone remodeling characterized by increased bone resorption and increased formation. It usually presents after the fourth decade of life and its prevalence increases with age to affect 1% to 2% of individuals over age 60 in the United States with a slight male predominance. There also are juvenile forms of the disease, often inherited, and familial cases account for approximately 15% of patients. PD has a greater prevalence in England and Northern Europe and is rare in Asian populations. It is usually a focal disease involving one bone (monostotic), but it can involve several bones (polyostotic). In rare cases, PD can become generalized, especially in familial disease. Osteogenic sarcoma is the dreaded complication of PD.

Etiology and Pathogenesis

Increased and abnormal osteoclastic bone resorption is the initiating event in PD. There is accumulating evidence that a slow virus plays a role in pathogenesis; candidates include:

- Paramyxovirus
- Respiratory syncitcal virus
- Measles virus
- Canine distemper virus.

Increased interleukin 6 (IL-6) production may be a contributing factor to the development of PD. Genetic mechanism are also likely to be important in familial cases, with possible linkage to chromosome 6 or 18.

The rate of bone resorption in PD can be increased over normal by as much as 10- to 20-fold. This is reflected in increased biochemical indices of bone resorption, including urinary excretion of collagen catabolites, like hydroxyproline and collagen cross-linked peptides, and osteoclast products, notably serum acid phosphatase (Table 1.8). Osteoblastic new bone formation responds appropriately to the increased resorption, and this is reflected in increased osteoblast products such as alkaline phosphatase and osteocalcin (Table 1.7). The increased cellular activity produces highly vascular and cellular bone. Osteoclasts can have up to 100 nuclei per cell, and collagen is deposited by overworked osteoblasts in an abnormal and disorganized pattern of "woven bone."

Because the tight coupling of formation and resorption is maintained in PD despite the increased skeletal turnover, systemic mineral homeostasis is usually unperturbed and serum calcium is usually normal. However, when patients with extensive PD are immobilized, as can occur upon fracture or surgery, they may become hypercalciuric or even hypercalcemic. This occurs because muscle stimulation of bone formation is decreased and bone resorption proceeds relatively unopposed. Hypercalcemia may also signal the presence of hyperparathyroidism, which is reported with an increased incidence in some PD series.

Symptoms and Signs

Paget's disease may affect any of the bones (Table 8.1), but the most common sites are the:

- Sacrum and spine (50%)
- Femur (46%)
- Skull (28%)
- Pelvis (60%).

TABLE 8.1 — SKELETAL INVOLVEMENT IN PAGET'S DISEASE
Pelvis
Spine
Sacrum
Femora
Skull
Scapula
Extremities
The involved bones are shown in generally decreasing order of involvement.

The clinical features of PD are:

- Bone pain
- Fractures
- Deformity
- Metabolic complications of the disease (Table 8.2).

However, many patients are asymptomatic (Table 8.3). Thus, PD is often discovered as an incidental radiologic finding or during the investigation of an elevated alkaline phosphatase level. Rarely, the disease presents with a pathological fracture. On physical examination, enlargement of the skull, frontal bossing, or deafness may be evident. Involvement of the weight-bearing long bones of the lower extremity often can result in bowing and an abnormal gait that contributes to the development of osteoarthritis in the hip and knee.

The early lesions of PD are often osteolytic, with (a) osteoporosis circumscripta and (b) pagetic lesions in the extremities that start in the metaphyses and move down the shaft as a V-shaped resorptive front. Osteoblastic lesions appear as PD evolves; such lesions can predominate in the late, "burned out" stages of this disease and have a sclerotic x-ray appearance.

TABLE 8.2 — SIGNS AND SYMPTOMS OF PAGET'S DISEASE
Skeletal • Pain • Deformity: weight-bearing bones • Fracture: weight-bearing bones • Secondary osteoarthritis: knees and hips • Malignancy: osteogenic sarcoma and other tumors
Neurologic (Compressive/Entrapment) • Headache: dizziness • Cranial nerves: eg, hearing loss • Craniospinal junction: hydrocephalus • Spinal cord and roots: – Paraplegia – Paresis • Peripheral nerves: specific entrapment syndromes • Back pain • Mental status change: platybasia
Other • Cardiovascular: high output • Metabolic: – Hypercalcemia – Hypercalciuria

TABLE 8.3 — CLINICAL PRESENTATION OF PAGET'S DISEASE
• Asymptomatic: 44% – Elevated serum alkaline phosphatase (16%) – Abnormal x-ray (28%) • Symptoms of Paget's disease: 56%

Thus, x-rays show a complex pattern of both radiolucency and radiodensity, depending upon the extent and stage of the disease. Bone scanning with technetium-labeled bisphosphonates or other bone-seeking agents is usually abnormal in active PD and can be used for surveying the skeleton. Bone scans are more sensitive than bone markers but too expensive for routine use. The increased blood flow to bone can cause high-output congestive heart failure, usually in patients with more than 50% involvement of the skeleton.

The most dreaded skeletal complication of PD is development of osteogenic sarcoma. The tumor may be difficult to diagnose early. High resolution MRI can help, but bone biopsy may be needed. In advanced stages, it is characterized by:

- Soft tissue swelling
- Increased pain
- A rapidly increasing alkaline phosphatase.

In addition to osteosarcoma, chondrosarcoma, and giant cell tumors occur in PD, with a combined incidence of about 1%.

Laboratory Findings

Markers of bone resorption and bone formation are increased in PD. Serum alkaline phosphatase (AP) is the most widely measured formation marker, and urinary hydroxyproline, an historic resorption marker. More specific immunochemical tests are now available, but they are not necessarily more sensitive than enzymatic AP (Tables 1.7 and 1.8). The basic biology of these bone markers is discussed in Chapter 1, *Skeletal Metabolism*, and illustrated in Figure 1.3.

For assessing resorption (Table 1.8), they include the urinary excretion products of type I collagen degradation and their different cross-linked molecules, including pyridinoline (PYD) and deoxypyridinoline (DPD). These include PYD or DPD and the crosslinks of C- and N-telopeptide collagen breakdown products, respectively named CTX and NTX.

Formation markers (Table 1.7) reflecting osteoblast activity are bone-specific alkaline phosphatase (BAP), osteocalcin (OC, BGP), and procollagen peptides (PICP, PINP). The serum calcium and phosphorus and the urinary calcium are usually normal in PD, but if there is immobilization, hypercalcemia and hypercalciuria can intervene.

■ Neurologic Manifestations

Neurological structures encased by or impinged on by the abnormal skeleton are at risk of compression or entrapment. Thus, the brain, cranial nerves, spinal cord, and peripheral nerves are all at risk. Sensorineural deafness occurs in up to 50% of patients in whom the skull is involved, and compression of the other cranial nerves can also occur. Involvement at the craniospinal junction can produce brain stem compression, obstructive hypocephalus, or vertebrobasilar insufficiency. Peripheral nerve entrapment syndromes include carpal and tarsal tunnel syndromes. Osteoarthritis is common in PD, especially in the knees and hips.

Diagnosis

The diagnosis of PD is made on the basis of:

- Characteristic clinical manifestations
- Serum and urinary bone markers
- Bone x-rays and scans.

The latter can also be used to document the extent and activity of disease. Bone biopsy is seldom necessary for diagnosis. PD can sometimes be confused with osteoblastic bone metastases, especially of breast and prostate carcinoma. Total AP has long been used for diagnosis and monitoring therapy. Although more specific bone markers have been recently developed (Chapter 1, *Skeletal Metabolism* and Chapter 10, *Osteoporosis*), they correlate well with AP and do not seem to have an advantage over this less expensive test.

Treatment

Until recently, treatment of PD was reserved for symptomatic patients. With the advent of several drugs that are effective in PD, indications for treatment (Table 8.4) have become more liberal and include:

- Bone pain
- Hypercalciuria and hypercalcemia
- Symptomatic heart disease
- Neurological involvement.

Furthermore, some clinicians advocate treating asymptomatic patients to inhibit the progression of disease. Pretreatment has also been advocated for orthopedic surgery in order to prevent excessive bleeding and postoperative hypercalciuria and hypercalcemia. Two classes of agents are used in the treatment of PD:

- Calcitonins (CTs)
- Bisphosphonates.

Plicamycin (Mithracin) is no longer recommended because of its toxicity. Cost may be a factor, depending on insurance coverage.

Asymptomatic patients with active PD involving sites that may impose a risk of fracture should be treated. In addition to relief of symptoms, the goal of treatment should be suppression of bone turnover markers into the normal range if possible, and retreatment when bone markers again rise above normal. This usually means treatment for several months followed by a period of vacation, with retreatment when markers again exceed normal. X-rays and bone scans may return toward normal after prolonged treatment.

TABLE 8.4 — DRUGS FOR THE TREATMENT OF PAGET'S DISEASE

Generic/*Trade* Name	**Manufacturer**	**Available Preparation**	**Recommended Dosage**
Alendronate/*Fosamax*	Merck	Tablets: 5, 10, 40 mg	40 mg/day × 6 months
Calcitonin			
Calcimar	Rhône-Poulenc Rorer	Injection: 2-mL vial, 200 IU/mL	50-100 IU × 6-18 months
Miacalcin	Novartis	Injection: 2-mL vial, 200 IU/mL	50-100 IU × 6-18 months
Etidronate/*Didronel*	Procter & Gamble	Tablets: 200, 400 mg	200-400 mg/day × 6 months
Pamidronate/*Aredia*	Novartis	Infusion: 30-, 60-, 90-mg vial	30 mg as 4 hour IV × 3 consecutive days
Risedronate/*Actonel*	Procter & Gamble	Tablets: 30 mg	30 mg/d × 2 months
Tiludronate/*Skelid*	Sanofi	Tablets: 200 mg	400 mg/day × 3 months

Abbreviations: IU, international unit; IV, intravenous (infusion).

See text for full discussion.

■ Calcitonin

Calcitonin acts to inhibit the abnormal bone resorption of PD by directly inhibiting osteoclast function. Both human and salmon calcitonin (HCT and SCT) can be used, with salmon predominating. SCT is administered parenterally or nasally, initially at a dose of 50 to 100 IU daily until symptoms are improved; thereafter, patients may be maintained on a less intense regimen such as 50 units 3 times a week. Improvement in pain is a prominent effect of SCT and is usually evident within 2 to 6 weeks. Serum AP and urinary collagen catabolites will fall by 50% within 3 to 6 months, with the AP lagging slightly behind.

Many patients will have a sustained response to treatment extending over years, and the biochemical parameters are often suppressed for 6 months to 1 year after treatment is discontinued. SCT is generally safe, although some patients will experience nausea and flushing. These may disappear upon continued treatment or become less disturbing if the drug is given before sleep. Up to 20% of patients receiving SCT treatment will develop resistance to it, which may be antibody-mediated. HCT therapy was effective in these patients, since salmon antibodies seldom bind HCT, but HCT is not readily available in the United States.

■ Bisphosphonates

Bisphosphonates are a group of drugs that inhibit osteoclast activity and number and become incorporated in bone matrix, perhaps thereby stabilizing it. The bisphosphonate etidronate disodium (Didronel) remains useful in PD. About 60% of patients will respond when it is administered in a dose of 5 mg/kg/d for 6 months. Several newer bisphosphonates appear to be superior to etidronate.

Alendronate (Fosamax) can be administered orally in a dose of 40 mg daily for 6 months. (This dose is higher than the 10 mg doses recommended for osteoporosis treatment and the 5 mg dose for osteoporosis prevention.) Alkaline phosphatase is suppressed by up to 80%, and biochemical remissions are often prolonged for more than 1 year after the drug is stopped. The main side effects are gastrointestinal (GI), sometimes requiring discontinuation of therapy in a small percentage of patients. This side effect can be minimized by instructing patients to take the drug in the morning with only water and to remain upright and fasting for the next one half hour. There appear to be no bone mineralization problems with alendronate, although this may occur with high doses of etidronate. Normal patterns of bone formation result when the rapid bone remodeling at pagetic sites was arrested. While normal indices of bone turnover are the goal of treatment, they are not always possible to achieve, so near-normal values may be more realistic.

It may be more convenient to administer an equivalent dose of a bisphosphonate intravenously. Pamidronate (Aredia) has been used for this purpose, although in the United States it is primarily for treatment of hypercalcemia. Pamidronate can be administered in several protocols ranging from daily to weekly or monthly infusions 30 to 90 mg, using higher doses in the more active disease. However, the Food and Drug Administration (FDA)-approved protocol allows 3 consecutive days of a 4-hour infusion of 30 mg daily, 90 mg total. About 30% of patients will experience flu-like symptoms after the first (and sometimes the second) infusion, with 24 hours of fever, myalgias, and headache.

Recently, tiludronate (Skelid), another bisphosphonate, was approved by the FDA for the treatment of PD as a 400-mg oral dose given for 3 months. It may be better tolerated than other bisphosphonates, but further studies are needed. Preliminary studies with yet another recently approved bisphosphonate, risedronate (Actonel), also appear promising. Both of these agents seem to have fewer GI side effects than alendronate, but further experience with them is needed to assess their GI effects. Patients receiving the bisphosphonates will benefit from calcium and vitamin D supplements to avoid hypocalcemia.

■ Other Therapies

The pain of PD responds to specific treatments, especially SCT. However, other common analgesics such as acetaminophen, aspirin and nonsteroidal anti-inflammatory drugs (NSAIDs) may also give relief, especially when there is a complicating osteoarthritis. Surgery can be used for specific indications. These include neurosurgery for the various compression syndromes and orthopedic surgery for deformities and hip replacement. Pretreatment several weeks prior to surgery with bisphosphonates can decrease bleeding and attenuate immobilization-induced hypercalcemia without apparently interfering with bone healing. Improved and early medical interventions with effective drugs should decrease the need for surgical intervention.

Summary

Paget's disease (PD) is a disorder of bone remodeling due to increased osteoclast-mediated bone resorption and abnormal compensatory bone formation. It usually occurs with a slight male predominance in adults, often of English or Northern European ancestry. But there are juvenile and familial forms of the disease that may provide clues to a viral and/or genetic etiology.

Paget's disease is usually a focal disease involving few bones, but it can also be generalized. Most bones can be involved, with the pelvis and spine being common targets. Bone pain is common, but PD can be asymptomatic, especially in its early course. Osteosarcoma is a dreaded but rare complication.

The abnormal remodeling can cause structural abnormalities of involved bone, such as bowing of the extremities and an enlarged skull. The abnormal bone can also encroach on adjacent nerves and cause complications such as hearing impairment and spinal cord impingement.

The biochemical hallmark of PD is an increased serum alkaline phosphatase, and other routine tests, including serum calcium and phosphorus, are usually normal; this distinguishes PD from osteomalacia, osteoporosis, and osteitis fibrosa cystica. Bone scan is more sensitive and practical than x-ray in documenting the extent of disease.

The availability of effective treatments for PD has revolutionized and liberalized clinical management. Treatment should be instituted for patients who are symptomatic or at risk for the complications of PD and when the serum alkaline phosphatase (AP) is 2-fold increased above normal. Furthermore, even asymptomatic patients, previously observed only, can be treated to prevent complications, as can candidates for bone surgery.

There are two groups of drugs that can be used for treatment, the calcitonins (CTs), usually salmon calcitonin (SCT), and the bisphosphonates; the latter are generally more effective. Several preparations of each are available, and new bisphosphonates with improved therapeutic indices are becoming available.

Most FDA-approved protocols recommend several months of treatment followed by a holiday. The AP should decline at least by 50% and preferably approach normal within several months. This approach may be most advantageous for newer bisphosphonates such as ibandronate and zolendronate that can be given by intravenous (IV) push in a single dose and produce months of therapeutic effect.

Orthopedic surgery can be used for specific indications and neurosurgery may be necessary for nerve impingement syndromes. Analgesics still have a role in pain management, although the bisphosphonates and especially SCT can relieve pain.

SUGGESTED READINGS

Medical Letter. 1997;39:65-66.

Chappard D, Retailleau-Gaborit N, Filmon R, Audran M, Basle MF. Increased nucleolar organizer regions in osteoclast nuclei of Paget's bone disease. *Bone*. 1998;22:45-49.

Cody JD, Singer FR, Roodman GD, et al. Genetic linkage of Paget disease of the bone to chromosome 18q. *Am J Hum Genet*. 1997;61:1117-1122.

Deftos LJ, Parthemore JG, Price PA. Changes in plasma bone GLA protein during treatment of bone disease. *Calcif Tissue Int*. 1982;34:121-124.

Deftos LJ, Wolfert RL, Hill CS. Bone alkaline phosphatase in Paget's disease. *Horm Metab Res*. 1991;23:559-561.

Garnero P, Gineyts E, Schaffer AV, Seaman J, Delmas PD. Measurement of urinary excretion of nonisomerized and beta-isomerized forms of type I collagen breakdown products to monitor the effects of the bisphosphonate zoledronate in Paget's disease. *Arthritis Rheum*. 1998;41:354-360.

Mee AP, Dixon JA, Hoyland JA, Davies M, Selby PL, Mawer EB. Detection of canine distemper virus in 100% of Paget's disease samples by in situ reverse transcriptase-polymerase chain reaction. *Bone*. 1998;23:171-175.

Parthemore JG, Burton DW, Deftos LJ. Associations and disassociations between GLA protein and alkaline phosphatase in skeletal metabolism. *J Orthop Res*. 1993;11:671-676.

Singer FR, Brown JP, Siris ES, et al. Integrated analysis of effectiveness of risedronate for Paget's disease. *Arthritis Rheum*. 1998;40:S279.

Singer FR, Clemens TL, Eusebio RA, Bekker PJ. Risedronate, a highly effective oral agent in the treatment of patients with severe Paget's disease. *J Clin Endocrinol Metab.* 1998;83:1906-1910.

Siris ES. Management of Paget's disease of bone in the era of new and more potent bisphosphonates. *Endocr Pract.* 1977;3:264-266.

Siris ES. Paget's disease of bone. *J Bone Miner Res.*1998;13:1061-1065.

Siris ES, Chines AA, Altman RD, et al. Risedronate in the treatment of Paget's disease of bone: an open label, multicenter study. *J Bone Miner Res.* 1998;13:1032-1038.

Siris E, Weinstein RS, Altman R, et al. Comparative study of alendronate versus etidronate for the treatment of Paget's disease of bone. *J Clin Endocrinol Metab.* 1996;81:961-967.

9 Renal Osteodystrophy

Introduction

Renal osteodystrophy (RO) is the complex skeletal disease of chronic renal failure, a bone disease that has several components:

- Osteitis fibrosa cystica, the disease of parathyroid hormone (PTH) excess
- Osteomalacia, the disease of vitamin D deficiency
- Extraskeletal calcification, and sometimes ossification, a disorder of the abnormal calcium and phosphorous metabolism of renal failure
- Adynamic bone disease, contributed to by PTH suppression.

In addition, osteosclerosis, osteoporosis, osteonecrosis, and, in the young patient, skeletal growth retardation commonly complicate renal failure (Table 9.1). 9

Etiology and Pathogenesis

The complex pathways involved in the development of RO are outlined in Figure 9.1. The failing kidney causes two major initiating events:

- Increased serum phosphorus
- Decreased 1,25-dihydroxyvitamin D (1,25-D) production by the kidney.

As outlined in Figure 9.1, the increased serum phosphorus (hyperphosphoremia) ultimately results in secondary hyperparathyroidism and osteitis fibrosia cystica. The decreased 1,25-D causes hypocalcemia

TABLE 9.1 — COMPONENTS OF RENAL OSTEODYSTROPHY
• Osteomalacia • Osteitis fibrosa cystica • Extraskeletal calcification • Adynamic bone disease

and osteomalacia. There are complex interactions between these two pathways such that essentially all of the primary calcemic hormones — PTH, vitamin D, and calcitonin (CT) — participate in pathogenesis.

■ Osteomalacia

The destruction of renal mass results in a decreased production by the kidney of 1,25-D. This deficiency of the active metabolite of vitamin D contributes to the development of the unmineralized bone matrix of osteomalacia. Other factors such as aluminum toxicity can also produce a mineralization disorder.

■ Osteitis Fibrosa Cystica Due to Secondary Hyperparathyroidism

Hypocalcemia due to the decreased gastrointestinal (GI) absorption of calcium caused by the vitamin D defiency and by the renal phosphate retention cause secondary hyperparathyroidism. The PTH excess causes osteitis fibrosa cystica. Paradoxically, the effect of the increased PTH is blunted by skeletal resistance to its actions due in part to the vitamin D deficiency.

The parathyroid gland becomes less responsive to its hormonal regulators because of decreased expression of its calcium and vitamin D receptors. Effective treatment to reverse these receptor changes requires both calcium and vitamin D replacement. The gland also undergoes monoclonal expansion as it be-

FIGURE 9.1 — PATHOGENESIS OF RENAL OSTEODYSTROPHY

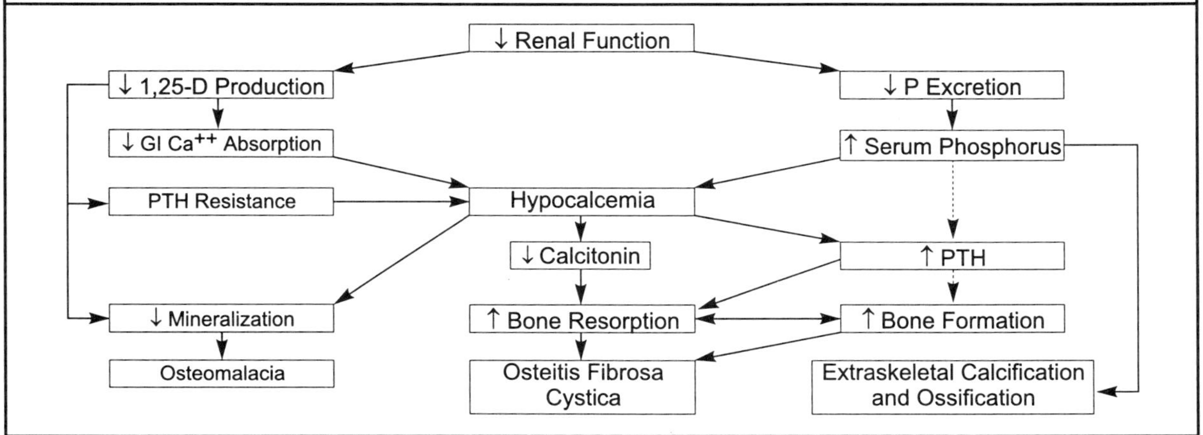

Abbreviations: 1,25-D, 1,25-dihydroxyvitamin D; Ca^{++}, calcium; GI, gastrointestinal; PTH, parathyroid hormone.

Dotted-line arrows indicate mechanisms that are not well established.

9

comes hyperplastic. While calcium regulates PTH secretion by the parathyroid gland in renal disease, the increased serum phosphate concentration further increases hormone biosynthesis.

■ Tertiary Hyperparathyroidism

Tertiary hyperparathyroidism is the name applied when the secondarily hyperplastic pararathyroid glands of renal failure escape from secretory control of PTH by calcium, secrete increased PTH concentrations, and thereby produce hypercalcemia. However, frank hypercalcemia is rarely a parathyroid excess-mediated event in renal failure. If hypercalcemia occurs, it is usually due to excess calcium and/or vitamin D administration. So the diagnosis of tertiary hyperparathyroidism should not be made unless hypercalcemia can be demonstrated after the discontinuation of vitamin D and calcium administration. This distinction is important because tertiary hyperparathyroidism, a monoclonal expansion of abnormal parathyroid cells, usually requires parathyroidectomy (PTX). In addition to calcium and vitamin D, other causes of hypercalcemia (Chapter 3, *Hypercalcemia*) should also be ruled out before the diagnosis of tertiary hyperparathyroidism is made.

■ Extraskeletal Calcification

The increased serum phosphorus causes the precipitation of calcium phosphate complexes at both skeletal and extraskeletal sites. Several clinical types are especially important:

- Calcification of medium-sized arteries
- Articular calcification
- Visceral calcification of the kidney, heart, and lungs
- Cutaneous calcification.

■ Adynamic Renal Osteodystrophy

This form of low turnover skeletal disease can be a significant problem in RO. Initially attributed to deposition of aluminum from drugs (eg, aluminum hydroxide) or dialysate fluid, adynamic bone disease can occur in dialysis patients who were never exposed to excessive amounts of aluminum from either aluminum-containing phosphate binders or dialysate aluminum. Iron deposits may play a role in some patients. Since PTH is lower in patients with adynamic bone disease than in other forms of RO, risk factors include those that will suppress PTH, such as the use of dialysis fluid with high calcium concentration, use of calcium-containing phosphate binders (eg, calcium carbonate), as well as the now rare use of aluminum-containing drugs.

Symptoms and Signs

Although the biochemical features of RO can be variable, in the early stages, progression of renal failure leads to an increase in serum phosphorus, a decrease in serum calcium and 1,25-D, and an increase of serum PTH. Serum CT is usually elevated but, like PTH, it may have decreased biological activity. Patients with RO have bone pain and tenderness, pathologic fractures, and muscle weakness. Extraskeletal calcium deposition causes vascular calcification, periarticular calcification, cutaneous calcification with uremia pruritus, and calciphylaxis, resulting in digital gangrene. Although markers of both bone formation and resorption (Chapter 1, *Skeletal Metabolism*) are increased, the effect of the abnormal renal function on their serum concentrations complicates clinical interpretation.

X-rays show signs of the osteomalacia (see Chapter 7, *Osteomalacia and Rickets*), secondary hyperparathyroidism (see Chapter 4, *Primary Hyperparathyroidism*), and osteosclerosis, often occurring near the end plates of adjacent vertebrate to produce a striped appearance — the "rugger jersey spine." Vascular calcification may also be evident on x-ray (Table 9.2).

TABLE 9.2 — BIOCHEMICAL AND CLINICAL FEATURES OF RENAL OSTEODYSTROPHY
Biochemical Features • Hyperphosphatemia • Hypocalcemia • Decreased serum vitamin D levels, notably 1,25-dihydroxyvitamin D (1,25-D) • Increased serum parathyroid hormone (PTH) • Increased bone turnover markers, notably alkaline phosphatase (AP)
Clinical Features • Skeletal: – Bone pain – Skeletal deformity – Growth retardation – Fractures • Extraskeletal complications: – Extraskeletal calcification – Calciphylaxis – Pruritus

Diagnosis

The diagnosis of RO is readily made since it invariably complicates chronic renal failure, especially end-stage disease. In addition to the abnormal renal function, characteristic blood tests include:

- Decreased serum calcium
- Increased serum phosphorus
- Increased serum alkaline phosphatase
- Decreased serum 1,25-D.

X-rays reveal the presence of the osteomalacia, the hyperparathyroidism, and extraskeletal calcification.

Treatment

The mainstays of treatment are vitamin D and calcium supplementation and dietary phosphate restriction. Calcitriol is the drug of choice, although the analog dihydrotachysterol (DHT) can also be used (Table 9.3). Calcitriol can also be given intravenously (0.5 to 3 μg) 3 times weekly in order to suppress PTH secretion without substantial hypercalcemia. In addition to restriction of dietary phosphate, often difficult to do, agents are given to bind dietary phosphate in their GI tract and prevent its absorption. Calcium, usually the carbonate salt, also binds phosphate and serves this purpose. Compared with calcium carbonate, calcium acetate seems to provide equivalent phosphate binding at about one half the calcium intake. The decrease in serum phosphate, in addition to increasing serum calcium directly, also inhibits PTH biosynthesis. If necessary, aluminum hydroxide preparations can be added to improve phosphate binding. However, these are now second line agents that should be used in moderation because of the recognized skel- 9

TABLE 9.3 — VITAMIN D FOR RENAL OSTEODYSTROPHY

Preparation	≈ Daily Dose
Vitamin D_2 or D_3	10,000 to 200,000 IU/day (0.25 to 5.0 mg/day)
Dihydrotachysterol (DHT)	0.20 to 2.4 mg/day
Caldifediol (25-OHD_3)	25 to 100 μg/day
Calcitriol	Oral: 0.25 to 1.0 μg/day;
Calcitriol (Calcijex)	IV: 1 to 3 μg 3 times/week

etal toxicities of aluminum. Desferrioxamine infusions (1.3 to 3.0 g, once weekly) can be used for aluminum chelation if indicated.

Patients who are frankly hypercalcemic or who have uremic pruritus or calciphylaxis may require subtotal PTX. However, as discussed above, tertiary hyperparathyroidism is rare. Calcimimetic agents, now experimental, are close to clinical application in treating the hyperparathyroidism of renal failure (see Chapter 2, *Calcium Metabolism*).

Clinical guidelines for the management of RO relate to control of serum phosphate, calcium, and vitamin D. There should be adequate calcium intake, usually 1 to 2 g of elemental calcium/day. The dialysate should provide adequate calcium. Every attempt should be made to control the serum phosphorus to 4.5 to 5.5 mg/dL. This usually requires the restriction of dietary phosphorus intake to 0.7 to 0.9 g/day and the use of phosphate binders such as calcium carbonate or calcium acetate. As discussed above, the use of aluminum hydroxide or aluminum carbonate for this purpose should be restricted. Vitamin D treatment is pivotal for RO, both as prophylaxis and treatment. There are several regimens that can be used (Table 9.3). Calcitriol is the drug of choice, and it can be given orally or parenterally at dialysis. Bisphosphonates are not yet used in renal osteodystrophy because of concerns about toxicity.

With effective medical therapy, PTX is seldom indicated in RO. Although tertiary hyperparathyroidism can benefit from PTX and gland autotransplantation, it is overdiagnosed. There must be persistent hypercalcemia predialysis and in the absence of vitamin D and calcium administration in order to make the diagnosis of tertiary hyperparathyroidism. Some other indications for PTX are pruritus unresponsive to other treatments and calciphylaxis with ischemic ulcers and necrosis. However, surgical removal of the parathyroid glands with or without autotransplantation make treatment of a difficult disease even more difficult.

Summary

Renal osteodystrophy (RO) is the name given to the complex of the skeletal disorders that occur in renal failure. There are four components to RO:

- Osteomalacia
- Osteitis fibrosa cystica
- Extraskeletal calcification
- Adynamic bone disease.

There are two abnormalities of the declining renal function that initiate this complex of skeletal disease:

- Increased serum phosphorus
- Decreased renal production of the active vitamin D metabolite, 1,25-D.

The increased serum phosphorus causes hypocalcemia, which in turn increases PTH secretion and, through this secondary hyperparathyroidism, causes the disease of PTH excess, osteitis fibrosa cystica. The decreased renal production of 1,25-D contributes to the hypocalcemia and causes the osteomalacia. The increased serum phosphorus also contributes to the hypocalcemia and causes the extraskeletal calcification. The pathogenesis of the adynamic bone disease is obscure. There are additional interactions of these disordered pathways that contribute to pathogenesis (Figure 9.1).

The symptoms and signs are those of renal failure and each of the skeletal abnormalities. Tests of renal function are abnormal, and there is an increased serum phosphorus, decreased serum calcium, and decreased serum 1,25-D.

In addition to treating the renal failure, treatment of RO is directed at the two initiating events:

- To decrease the serum phosphorus
- To increase the serum 1,25-D.

Phosphate binders, a low phosphorus diet as acceptable, and dialysis are used to treat the former. Vitamin D metabolites are used to treat the latter. Calcitriol is the preferred drug since it represents the renal metabolite whose production is impaired. It can be given orally or parenterally, the latter usually at the time of dialysis. However, other vitamin D metabolites may be used. Adequate calcium is provided by dietary supplements and by the dialysate calcium for the dialyzed patient. Parathyroidectomy (PTX) should be reserved for those few patients that can be conclusively diagnosed as having tertiary hyperparathyroidism. Renal transplantation is the ultimate treatment.

SUGGESTED READING

Antonsen JE, Sherrard DJ, Andress DL. A calcimimetic agent acutely suppresses parathyroid hormone levels in patients with chronic renal failure. Rapid communication. *Kidney Int.* 1998;53:223-227.

Argiles A, Mourad G. How do we have to use the calcium in the dialysate to optimize the management of secondary hyperparathyroidism? *Nephrol Dial Transplant.* 1998;13:62-64.

Arnold A, Brown MF, Urena P, Gaz RD, Sarfati E, Drueke TB. Monoclonality of parathyroid tumors in chronic renal failure and in primary parathyroid hyperplasia. *J Clin Invest.* 1995;95:2047-2053.

Arnold A, Stewart AF. The molecular basis of disorders of the parathyroid gland. In: Jameson JL, ed. *Textbook of Molecular Medicine.* Cambridge, Mass: Blackwell Scientific. In press.

Bouillon R. The many faces of rickets. *N Engl J Med.* 1998;338: 681-682. Editorial.

Duerksen DR, Papineau N. Electrolyte abnormalities in patients with chronic renal failure receiving parenteral nutrition. *J Parenter Enteral Nutr.* 1998;22:102-104.

Frazao JM, Chesney RW, Coburn JW, 1 Alpha D 2 Study Group. Intermittent oral 1-alpha-hydroxyvitamin D_2 is effective and safe for the suppression of secondary hyperparathyroidism in haemodialysis patients. *Nephrol Dial Transplant.* 1998;13:68-72.

Goicoechea M, Vazquez MI, Ruiz MA, Gomez-Campdera F, Perez-Garcia R, Valderrabano F. Intravenous calcitriol improves anaemia and reduces the need for erythropoietin in haemodialysis patients. *Nephron.* 1998;78:23-27.

Llach F, Yudd M. The importance of hyperphosphataemia in the severity of hyperparathyroidism and its treatment in patients with chronic renal failure. *Nephrol Dial Transplant.* 1998;13:57-61.

Messa P, Mioni G, Turrin D, Guerra UP. The calcitonin-calcium relation curve and calcitonin secretory parameters in renal patients with variable degrees of renal function. *Nephrol Dial Transplant.* 1995;10:2259-2265.

Olaizola I, Aznarez A, Jorgetti V, et al. Are there any differences in the parathyroid response in the different types of renal osteodystrophy? *Nephrol Dial Transplant.* 1998;13:15-18.

Ott SM. Calcimimetics – new drugs with the potential to control hyperparathyroidism. *J Clin Endocrinol Metab.* 1998;83:1080-1082. Editorial.

Rufino M, De Bonis E, Martin M, et al. Is it possible to control hyperphosphataemia with diet, without inducing protein malnutrition? *Nephrol Dial Transplant.* 1998;13:65-67.

Slatopolsky E. The role of calcium, phosphorus and vitamin D metabolism in the development of secondary hyperparathyroidism. *Nephrol Dial Transplant.* 1998;13:3-8.

Slatopolsky E, Delmez JA. Pathogenesis of secondary hyperparathyroidism. *Miner Electrolyte Metab.* 1995;21:91-96.

Velásquez FF, Altamirano E, Ramos PT. High frequency of iron bone deposits in a Mexican population with renal osteodystrophy. *Nephrol, Dial, Transplantation.* 1998;3(suppl 3):46-50.

10 Osteoporosis

Introduction

Osteoporosis is a systemic skeletal disease characterized by increased porosity of the skeleton leading to decreased bone mass and a consequent increased risk of fragility fractures. The most common sites of fracture are:

- Spine
- Hip
- Wrist – Colle's fracture.

Osteoporosis is the most common metabolic bone disease, effecting an estimated 75 million people, primarily postmenopausal females, in Europe, the United States, and Japan. In the United States, approximately half of postmenopausal Caucasian women have low bone mass and another 25% have osteoporosis; there are approximately 1.5 million new fractures per year. Fifteen percent of females develop osteoporosis during the sixth decade. Osteoporosis is most common in Caucasians and least common in Blacks. The highest rate of fractures is also in Caucasians, followed by Asians, Hispanics, and Blacks, but significant risk is present in all ethnic groups.

Osteoporosis can be defined by an intermediate outcome – low bone mineral density (BMD) (see Chapter 1, *Skeletal Metabolism*). The World Health Organization (WHO) recommends definitions based on T-scores (standard deviation [SD] from the mean for young adult Caucasian women) (Table 10.1). The T-score is used because, as discussed subsequently, changes in it reflect fracture risk. The remaining life-

TABLE 10.1 — WHO DEFINITIONS BASED ON BONE MINERAL DENSITY	
Classification	**T-Score**
Normal	Above -1
Osteopenia (low bone mass)	-1.0 to -2.5
Osteoporosis	-2.5 and below
Abbreviations: WHO, World Health Organization.	

time fracture probability (RLFP) progresses from 1.6 at a T-score of -2.5 to 5.4 at a T-score of -5.0. While these definitions can establish the prevalence of osteoporosis, they are not the sole determinants of management decisions. Patients with osteoporosis have increased fracture risk. Osteopenia (poor or less bone) refers to decreased bone mass in the absence of fracture. Patients with osteopenia do not have increased fracture risk, but may be at risk of developing osteoporosis and fracture risk in the future.

Based on bone mass measurements, osteoporosis is estimated to affect 30% of all postmenopausal Caucasian women in the United States, and the fraction affected rises with age. Site specific estimates are approximately 16% for the spine, 16% for the hip, and 17% for the mid-radius. The lifetime risk of hip fracture, the most devastating of the osteoporotic fractures, is 17% in Caucasian women and 6% in Caucasian men. Fractures lead to hospitalization and long-term nursing care with its attendant costs, and most fracture patients do not regain their independence.

Direct health-care costs associated with osteoporosis are approximately fourteen billion dollars annually in the United States, most being due to hospitalization and nursing-home care. Although the etiology of osteoporosis is not fully elucidated, many risk factors have been associated with the development of the disease, and a few of them approximate osteoporosis causation. There is no cure for osteoporosis, but there are many effective treatments. In many cases, osteoporosis may be preventable.

Osteoporosis is usually considered a disease of women in that 80% of those affected are women and one in two women over age 50 will have an osteoporosis-related fracture in her lifetime. However, the disease also affects males – 2 million American men have osteoporosis and one in eight men will have an osteoporosis-related fracture in his lifetime. In general, osteoporotic fractures in males begin to increase several decades later than in females. While the hip fracture rate is 2 to 3 times higher in women than men, the 1 year mortality following hip fracture is 2 times higher in men, and while the incidence of osteoporosis in Blacks is less – Black women over age 65 have half the fracture incidence of Caucasian women of the same age – 10% of Black women over age 50 have osteoporosis and an additional 30% have increased fracture risk due to low bone density.

Primary osteoporosis refers to the disease in the aging population where the onset of the disease can better be related to risk factors rather than causal, etiological factors. In this category are postmenopausal osteoporosis and the osteoporosis of aging, the latter sometimes referred to as senile osteoporosis. Postmenopausal osteoporosis is sometimes classified as type I and characterized as a loss of primarily trabecular bone due to the decline in estrogen production. Type II osteoporosis refers to age-related osteoporosis, which is characterized by a loss of both trabecular and cortical bone in both males and females. Rather than or in addition to estrogen deficiency, type II osteoporosis is postulated to be due to age-related declines in the function of organs maintaining calcium metabolism and secondary hyperparathyroidism. This classification remains controversial. A unitary model posits estrogen deficiency as a key cause of both types.

Secondary osteoporosis refers to bone loss associated with and probably caused by specific clinical disorders, such as hyperthyroidism and hyperadrenocorticoidism. The main emphasis of this chapter is primary osteoporosis.

Etiology and Pathogenesis

The development of osteoporosis results from a relative increase of bone resorption over bone formation in the adult skeleton (Figure 10.1). This uncoupling of bone metabolism can occur through an increase in bone resorption or a decrease in bone formation; both are likely to contribute. The resultant low bone mass does not have sufficient structural integrity to sustain the mechanical needs of the skeleton, so fracture results. The gross mechanical inadequacy of the skeleton leading to fracture is represented at the microscopic level by microarchitectural deterioration, including trabecular perforation and disconnection.

There is a direct relationship between bone mass, measured and quantitated as bone mineral density (BMD), and fracture risk. For a 1 SD decrease in BMD, which represents 10% to 12% decline in BMD, there is a 1.4- to 2.5-fold increased risk of fractures. This relationship between BMD and increased (relative) fracture risk is closest for the site of measurement, but a low BMD at one site is also associated with an increased relative risk of fracture at any skeletal site. Other risk factors magnify the fracture risk of decreased BMD. For example, it has been estimated that a personal or family history of fracture increases risk by about 2-fold and that age adds a 1.5-fold risk for every 5 years after peak bone mass. Thus, it is

FIGURE 10.1 — PEAK BONE MASS AND OSTEOPOROSIS

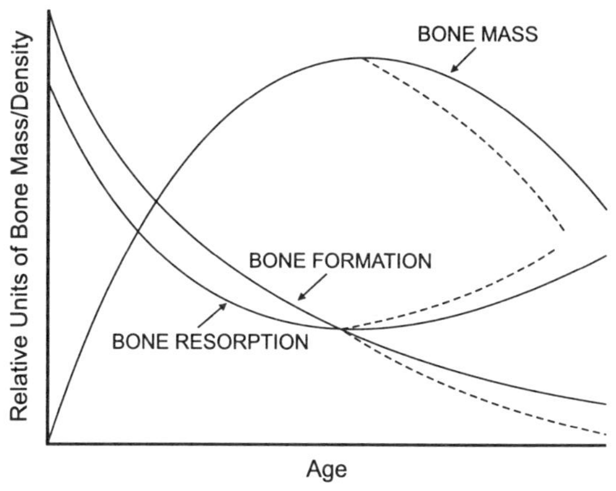

Schematic representation of the development of osteoporotic loss of bone mass/density. The solid lines represent normal skeletal physiology (see Figure 1.1). The dashed lines represent the increased loss of bone mass and consequent decrease in bone mineral density (BMD) that occur in osteoporosis due to increased bone resorption and/or decreased bone formation.

10

likely that other well documented risk factors contribute several percentage points to overall fracture risk.

Osteoporosis Risk Factors

Many such risk factors have been described for the development of osteoporosis. They are summarized in Table 10.2.

Most of the risk factors have been identified through epidemiological studies, so causality has not been clearly established. Nevertheless, they appear to contribute independently to fracture risk. It is im-

TABLE 10.2 — OSTEOPOROSIS RISK FACTORS
Low Bone Mineral Density (BMD)
Age and Gender
Gonadal Steroids (decreased lifetime exposure) • Late menarche • Early menopause • Oopherectomy, especially premenopausal • Amenorrhea • Nulliparity • Gonadal failure or loss • Drug-induced suppression, eg, glucocorticoids
Nutrition and Mineral Metabolism Factors • Low intake, especially during bone growth of: – Calcium – Vitamin D (or inadequate sunlight) • Chronic excess of: – Phosphorus and sodium – Thyroid hormones – Protein – Caffeine, cigarettes, alcohol – Calciuretic agents, especially thiazides • Defect in renal calcium conservation • Low body weight • Rapid and substantial (≥ 10%) weight loss • Anorexia nervosa and bulimia • Increased bone markers
Heredity/Genetic Factors • Race: Caucasian or Asian • Female gender and premature graying • Relative with osteoporosis • Relative with nontraumatic fracture • Lean habitus: low weight-to-height ratio • Longer hip axis length
Immobilization (chronic) • Spinal-cord injury • Arthritides • Stroke
Life-style Factors
Medications and Illness
Fracture in Patient or First-Degree Relative
Falling

portant to note that the risk factors identified in Table 10.2 are for both fracture and bone loss and usually both. It is also important to note that some risk factors are modifiable, such as low BMD, decreased gonadal steroids, poor nutrition and mineral metabolism, adverse lifestyle, and falling, while others, such as genetic factors and age, are not. The modifiable risk factors are the objects of therapy (Table 10.3).

TABLE 10.3 — MODIFIABLE RISK FACTORS IN OSTEOPOROSIS
• Low bone mineral density (BMD) • Decreased gonadal steroids • Low calcium and vitamin D intake • Low body weight and rapid weight loss • Excess smoking and alcohol consumption • Sedentary lifestyle • Falling • Chronic treatable diseases

Bone Mineral Density and Age: Peak Bone Mass

Age-related loss of bone mass is an important component of essentially all forms of osteoporosis. Peak bone mass is usually obtained in adults between the fourth and fifth decades (Figure 1.1). Factors that affect the peak bone mass include:

- Gender
- Race
- Heredity
- Nutrition
- Physical activity.

Men have a higher peak bone mass than women; Blacks have a higher peak bone mass than Caucasians or Asians. Dietary sufficiency of calcium and vitamin D, especially during adolescence, promote bone mass development, as does physical activity during this time. By about age 20, the average woman has acquired most of her skeletal mass. Following peak bone mass attainment in the fourth decade, there is a progressive decline in bone mass as a function of time in essentially all individuals. This decline is greater in females than in males, begins in the perimenopause, and is accelerated after the menopause. Women can lose up to 20% of their bone mass in the 5 to 7 years following the menopause. This increased rate of bone loss becomes attenuated before the decade after the menopause.

It can be difficult to draw a distinct line between the normal process of age-related loss of bone mass and the pathological process leading to the development of osteopenia and the disease osteoporosis (Figure 10.1). The occurrence of fracture is the clinical event that makes this distinction.

■ Gonadal Steroids

Sex hormones are important in both genders for the acquisition and maintenance of bone mass, and hypogonadism is associated with a decreased bone mass in both males and females. Estrogen deficiency is clearly associated with the development of osteoporosis. The advent of the menopause accelerates age-related loss of bone mass, with a greater loss in trabecular bone than cortical bone. By contrast, prolonged estrogen exposure, as occurs in early menarche and delayed menopause, and perhaps pregnancy and the use of birth control pills, helps to increase and sustain bone mass.

Estrogen deficiency can also be caused by amenorrhea and oophorectomy. The former may occur in female athletes who exercise to excess and in women with anorexia nervosa, bulimia, and pituitary disease, such as tumors. It can also occur when endometriosis is treated with luteinizing hormone releasing hormone (LHRH) agonists to induce artificial menopause.

The low BMD found in anorexia nervosa likely has a nutritional component. However, there are other factors, such as increased adrenal steroids and decreased gonadal steroids, that contribute to this complex disorder. Although a role for estrogen in the pathogenesis of osteoporosis is well-supported, it should be kept in mind that studies demonstrating a difference in measurable estrogen status between postmenopausal females with and without osteoporosis have been complex.

In postmenopausal females, there is generally a direct relationship between serum 17β-estradiol and BMD and an inverse relationship with bone loss and fracture risk at the hip and spine. However, higher serum levels of estrone are associated with increased fracture risk. Although estrone is not as potent as estradiol, its bioavailability may be comparable because of its high serum levels and lack of binding to sex hormone-binding globulin (SHBG). In fact, high serum SHBG also correlated with increased hip and vertebral fractures in postmenopausal females. In older men, circulating estrogens predict the rate of bone loss better than serum androgens, and boys with mutations in the estrogen receptor or the aromatase that converts androgens to estrogens have low BMD. A subclinical decrease in gonadotropins and gonadal steroids may interfere with the development and maintenance of bone mass. Bone cells can also metabolize gonadal steroids and thus locally regulate themselves.

The role of testosterone in sustaining the male skeleton during the aging process is not as clear. The development of an andropause in males remains controversial. However, in males after the eighth or ninth decade, when there may be a noticeable decline in serum testosterone, osteoporotic fractures increase in incidence and prevalence. Although the issue of an age-related decline in testosterone is controversial, other causes of testosterone deficiency are associated with a decrease in bone mass. These include hypogonadal disorders due to pituitary and testicular disease, such as Klinefelter's syndrome, and androgen deficiency states induced by treatments for prostate cancer, especially orchiectomy.

It has become recently appreciated that estrogens in males and androgens in females help to sustain skeletal mass. Females with androgen excess either endogenous, as may occur with certain gonadal and adrenal tumors, or exogenous, as may occur when androgens are taken for indications that have not been approved by the Food and Drug Administration (FDA), have increased bone mass. A further decrease in the low levels of estrogens in males, as occurs with certain rare genetic diseases of estrogen metabolism and the estrogen receptor, may lead to reduced bone mass. The best evidence for the importance of gonadal steroids in skeletal health is the beneficial effect that estrogen and androgen therapy respectively have in maintaining bone mass and increasing it in the hormone deficient patient.

■ Calcium-Regulating Hormones

In general, primary abnormalities in PTH, CT, or vitamin D are not of pathogenetic significance in osteoporosis. However, secondary changes in these hormones may contribute to pathogenesis. Thus, negative calcium balance may lead to secondary hyperparathyroidism and decreased gonadal steroids may reduce CT secretion. This family of hormones is important in the treatment of osteoporosis, as discussed later.

■ Nutrition — Calcium and Vitamin D

A role for dietary deficiency of calcium and vitamin D in the primary pathogenesis of osteoporosis is tempting to accept but remains to be proven. Although there are some classical studies to support such a role, recent studies have not been consistent. The disease usually associated with vitamin D and/or calcium deficiency is osteomalacia, not osteoporosis. Nevertheless, it seems prudent to recommend an adequate dietary calcium, especially during adolescence for females in whom dietary intake of calcium and vitamin D is so low. The deleterious skeletal effect of low calcium (and vitamin D) intake in adolescent females (who do not drink much milk anymore) is aggravated by their increased consumption of high phosphate diet drinks. Although some drinks are calcium supplemented, the bioavailability of the calcium in, for example, orange juice, has not been established for all preparations.

The recommended dietary allowance (RDA) for females as well as males is 1200 mg from age 11 to 24 and 800 mg from age 25 onward. A 1994 National Institutes of Health (NIH) consensus conference on calcium recommended that calcium intakes be 1000 mg for women beginning at the menopause and 1500 mg for postmenopausal women who are not receiving any form of estrogen replacement therapy. Similarly, most experts recommend a dietary intake of vitamin D of 600 to 800 units, well above the RDA recommendation of 200 to 400 units.

Although most Americans have sufficient sunlight exposure and/or vitamin D in their diets to assure adequate absorption of dietary calcium, certain populations do seem to be at risk for vitamin D and calcium deficiency. (It is ironic to note that vitamin D deficiency occurs in some Middle-Eastern cultures where there is adequate sunlight but where culturally-mandated clothing in females and over-swaddling of infants inhibits sunlight exposure.)

Elderly females who live at high latitudes, especially if they are institutionalized, can have vitamin D deficiency as reflected by low serum concentrations of 25-hydroxyvitamin D (25-D), less than 15 ng/dL.

Furthermore, there seems to be an age-dependent loss of the ability to absorb vitamin D, as well as calcium, and to metabolize the vitamin normally to its active forms. Thus it seems prudent to assure for the elderly, in addition to calcium supplementation, an adequate vitamin D intake. Other nutrients, especially sodium and protein, have been associated with an increased urinary calcium loss and the low bone mass. However, recent studies do not support this association. Of course, a diet adequate in all important nutrients should be adhered to.

In 1997, the National Academy of Sciences made nutritional recommendations (Table 10.4). They also set an upper limit for calcium of 2.5 g/day and for vitamin D of 2000 IU/day. While the preferred sources for calcium and vitamin D, it is unlikely that foods can meet requirements (the only vitamin D supplemented food in the United States is milk), so supplements are usually necessary (Tables 5.4 and 5.5)

TABLE 10.4 — NUTRITIONAL RECOMMENDATIONS OF THE NATIONAL ACADEMY OF SCIENCES

Age (Years)	**Calcium** (mg)	**Vitamin D** (IU/d)
19-50	1,000	200 (5 mcg)
51-70	1,200	400
> 70	1,200	600

■ Constitutional Factors

Body weight is positively correlated with bone mass, and the usual patient with osteoporosis is of slight body build. By contrast, osteoporosis is less common in the obese individual. Changes in body weight also correspondingly influence BMD. Postmenopausal women with more than a 50% weight gain since age 25 have a substantially lower incidence of hip fracture than women who have not gained such weight with increasing age. The usual explanation for this phenomenon is the beneficial effect as a stimulus for bone formation of weight bearing on the skeleton. Fat may also be an estrogen store. Substantial (≥ 10%) and rapid weight loss increases the risk of loss of BMD. Other factors that seem to have the same beneficial effect are body size and muscle strength. Many of these constitutional factors may have a genetic component. In fact, genetic factors have a predominant effect on peak bone mass.

■ Heredity

The importance of hereditary factors on the development of osteoporosis is becoming increasingly appreciated. Twins studies and mother-daughter comparisons indicate that approximately 60% to 80% of the variance in bone density is genetically determined. A maternal history of fracture doubles a woman's risk of fracture. The specific hereditary causes of osteoporosis have not been identified, but there is an intense search for genes that may play such a role. Although controversial, some studies suggest that polymorphism in the genes for the vitamin D receptor (VDR) and for type I collagen may predispose to the development of osteoporosis. Candidate genes include:

- VDR and type I collagen
- Estrogen receptor
- Interleukin-6 (IL-6)
- Insulin-like growth factor-I (IGF-1)
- Transforming growth factor-beta (TGF-β)
- Calcitonin (CT) receptor.

■ Lifestyle

Several factors that could be loosely grouped under lifestyle are reported to be associated with the development of osteoporosis. Many studies show a relationship between cigarette smoking and low bone mass in both genders. Low body weight has been suggested as the mechanism for this relationship, but studies on this point are conflicting. Alcohol abuse has been associated with low bone mass and fracture risk in both men and women. The mechanism of alcohol's effect may be a direct toxic action on osteoblasts or through indirect effects on the gonads and the liver. However, moderate alcohol intake may not be harmful and may even slightly reduce bone loss in women. Some studies have shown an association between caffeine intake and low bone density and increased fracture risk. A proposed mechanism for many of these factors has been increased urinary calcium excretion.

■ Exercise

Although exercise can be generally recommended for overall and skeletal heath, the precise role of the exercise in bone loss and osteoporosis remains controversial. Several studies posit a direct correlation between intensity of exercise and beneficial effects on bone mass. The greatest beneficial effect of the exercise on the skeleton seems to occur in childhood and adolescence, where exercise, especially high impact exercise, is important in achieving maximal peak bone mass. The value of intense exercise is also demonstrated in unilateral sports activities such as tennis, where the more active limb has greater bone density. However, if intense exercise in females causes menstrual irregularities, the consequent decrease in estrogen exposure is deleterious to the skeleton. In general, the effects of exercise in preventing bone loss in postmenopausal females are not substantial, and affects on fracture risk are not established. Physical activity may act to maintain the skeleton rather than increase BMD. Overall, however, some exercise seems better than none for skeletal health, but it must not be of the type that can increase fracture risk.

■ Medications and Illness

Several drugs have been implicated in the loss of bone mass. They include:

- Anticonvulsants
- Diuretics
- Thyroid hormones
- Glucocorticoids.

Studies of the first two are neither convincing nor consistent. Thiazide diuretics have been variably reported to increase bone mass by conserving renal calcium. Although meta-analysis of several studies found a decreased risk of fracture among thiazide users, there have been no randomized trials. Thyroid hormone excess, exogenous or endogenous, can lead to increased bone resorption. However, the clinical importance of such an occurrence has probably been overestimated. Thus, while steroid-induced osteoporosis is well documented and discussed in detail, subsequently, the evidence for a role of thyroid hormone in the pathogenesis of osteoporosis is less certain. Diseases that specifically affect the skeleton can contribute to the development of osteoporosis. Diseases with generalized effects can also have a deleterious effect on bone mass, such as:

- Renal disease
- Gastrointestinal (GI) disease
- Connective tissue disease
- Immobilization
- Malignancy
- Diabetes.

These issues are discussed in more detail under secondary osteoporosis.

Signs and Symptoms

Osteoporosis may develop over many years with few signs and symptoms. This feature along with its high incidence and prevalence has caused osteoporosis to be referred to as the silent epidemic. In fact, it has been estimated that fewer than 25% of patients with osteoporosis have been identified. As there are commonly no warning signs of osteoporosis, fracture is usually the presenting clinical manifestation of the disease. While hip and wrist fracture are dramatic events, vertebral fractures may be silent but eventually cause kyphosis and height loss (Table 10.5). Because of the silent nature of disease development, all postmenopausal women should be evaluated for the risk factors of osteoporosis.

TABLE 10.5 — SIGNS AND SYMPTOMS OF OSTEOPOROSIS
Asymptomatic – Common
Fractures • Vertebra • Hip • Wrist • Other
Bone Pain
Loss of Height
Kyphosis

Fracture is the most serious complication of osteoporosis. The lifetime fracture risk in 50-year-old Caucasian women is approximately 17% for femoral, 15% for vertebral, 16% for distal forearm, and 40% for any of the three. The comparable figures are lower for males: 6%, 5%, 3%, and 13%, respectively. All fractures are serious, but hip fracture carries with it the greatest morbidity and even mortality. There is a greater than 50% incidence of persistence of pain for vertebral fractures, and many patients are unable to resume their normal activities, even after vertebral fracture and especially after hip fracture. About one fourth lose their ability to ambulate and approximately 20% to 30% of patients with hip fracture die within 1 year. A lower but substantial number have a similar fate following vertebral fracture. Oral bone loss may cause tooth loss.

Laboratory Tests

Most patients with osteoporosis have normal routine laboratory tests, including serum calcium and phosphorus. In addition, most specialized laboratory tests are also normal, including serum parathyroid hormone (PTH) and vitamin D metabolites. Thus, the usually normal biochemistries of the patient with osteoporosis can be readily distinguished from the abnormal biochemistries of patients with other common metabolic bone diseases (Table 10.6). However, in some patients with osteoporosis there may be increased serum PTH, thought to be due to secondary hyperparathyroidism caused by calcium deficiency or loss; decreased 25-D, thought to be due to dietary deficiency; and decreased 1,25-dihydroxyvitamin D (1,25-D), thought to be due to decreased renal production.

■ X-ray Studies

Unless there is a characteristic fracture, routine x-rays are not helpful in establishing the diagnosis of osteoporosis. Approximately 30% to 50% of bone mass must be lost before the decreased bone mass of osteoporosis is apparent by routine x-ray. Furthermore, overexposure of the x-ray film may lead to the appearance of osteoporosis when it is not actually

TABLE 10.6 — ROUTINE LABORATORY TESTS IN THE DIFFERENTIAL DIAGNOSIS OF METABOLIC BONE DISEASES

Disease	Calcium	Phosphorus	Alkaline Phosphatase	PTH
Osteoporosis	Normal	Normal	Normal	Normal
Osteomalacia	↓	↓	↑	↑
Paget's disease	Normal	Normal	↑	Normal
Ostetitis fibrosa cystica (PH)	↑	↓	↑	↑

Abbreviations: PTH, parathyroid hormone; PH, primary hyperparathyroidism.

present. Nevertheless, radiographic evidence of osteopenia can be a strong predictor of low bone mass.

■ Bone Densitometry

Bone mineral density can be used to:

- Establish the diagnosis of osteoporosis
- Confirm the diagnosis in a patient with fragility fracture
- Predict future fracture risk
- Monitor the progress of the disease
- Monitor the effects of therapy.

The diagnosis of osteoporosis can only be made by a standardized procedure that directly and quantitatively measures bone mass and quantitates it (Table 10.7). Bone mass measurement is expressed as the bone mineral content (BMC, grams) or bone mineral density (BMD, grams/cm^2). The preferred contemporary technique for measurement is dual energy x-ray absorbtiometry (DXA) with bone mass expressed as BMD. Assessment is usually made at the hip and the lumbar spine. The proximal femur includes the femoral neck, the greater trochanter, the intertrochanter, and Ward's triangle. In most circumstances, the total hip is the most useful measurement and is the default plot of most software packages. DXA allows a rapid and precise method of assessing BMD and delivers a radiation dose less than that delivered by a routine chest x-ray. BMD measurement is indicated when the result could influence treatment. The National Osteoporosis Foundation (NOF) recommends BMD for:

- All postmenopausal women under age 65 who have at least one more risk factor for osteoporosis
- All women ≥ age 65
- All patients with osteoporotic fractures

TABLE 10.7 — BONE MASS MEASUREMENT

Method	Site	Comment
Dual energy x-ray absorptiometry (DXA)	Spine, hip, forearm, total body	High precision and accuracy; most widely used and standardized
Quantitative computed tomography (QCT)	Spine, forearm	Most radiation exposure; independent analysis of cortical and trabecular bone
Radiographic absorptiometry (RA)	Hand	Widely available; processed off-site
Magnetic resonance imaging (MRI)	Any site	Evaluates trabecular connectivity; interferometry
Single energy x-ray absorptiometry (SXA)	Forearm, heel	Not applicable to the axial skeleton
Ultrasonography (US)	Heel, fingers, tibia, patella	No radiation; portable; measures other parameters of bone strength

- Women considering hormone replacement therapy (HRT) for whom BMD would facilitate the decision
- Women on HRT for prolonged periods.

There is a direct correlation between BMD as assessed by DXA and fracture risk. BMD has an inverse relationship to the risk of fracture. For every SD decrease in BMD, fracture risk approximately doubles. This inverse relationship is generally logarithmic and continuous, so attempts to define a fracture threshold have been necessarily arbitrary. However, a therapy-related increase in BMD does not necessarily result in a reduced fracture risk, and some therapies can reduce fracture risk without causing a substantial increase in BMD. In fact, there are not data that show a relationship between the magnitude of therapy-induced changes in BMD and change in fracture risk. Thus, it is important to distinguish between BMD as a diagnostic procedure and as monitor of therapy. While the former generally correlates with fracture risk, therapy-related BMD changes as the latter may not. In fact, osteoporosis treatments can produce a substantial decrease in fracture rate without a substantial increase in BMD, and vice versa. Thus, factors other than BMD, such as the quality and strength of bone, also affect fracture rate.

Bone size must be considered along with BMD in the clinical assessment of the skeleton. Males do not have a greater BMD than females, but bigger bones. Similarly, most of the structural skeletal advantage that Blacks have over Caucasians seems to be due to bigger bones rather than greater BMD. In trabecular bone, BMD reflects trabecular number, trabecular thickness, and/or trabecular density. BMD precision and accuracy can be impaired by poorly maintained equipment, patient movement, and coexisting disorders, such as degenerative joint disease, spondylopathy, heterotopic ossification, vessel calcification, and vertebral fractures.

The World Health Organization (WHO) has recommended that the mean BMD of healthy young adults be considered as the reference level. The SD variations above and below these means are described as T-scores, and these scores are used to represent the individual being tested (see *Introduction*). Low bone mass, termed osteopenia, is defined as a BMD between 1.0 and 2.5 SD below this mean. Osteoporosis is defined as a BMD of 2.5 SD or more below this mean. Severe or established osteoporosis is defined as a BMD more than 2.5 SD below this mean, occurring in the presence of one or more fractures. As detailed subsequently, the NOF uses an osteopenia cutoff of -2 SD, based on a 98% confidence limit rather than the 95% limit used by the WHO, so more individuals will be classified as osteopenic.

While T-scores are now the standard, some test results also present a Z-score, which is the comparison to age-matched reference ranges. Although these latter comparisons were generally not as useful for assessing fracture risk in clinical practice, it has been observed that the risk of hip fracture is increased by 2- to 3-fold for each SD of BMD below the mean for age. Thus, a Z-score lower than -1 indicates the need to prevent further bone loss. The "worst" score is usually used for therapy decisions.

In general, osteopenic patients should be candidates for preventive therapy, while osteoporotic patients are obviously candidates for disease treatment. However, it should be kept in mind that BMD standards are based primarily on studies of Caucasian females. Where ethnic data are available, they also oversimplify by variably combining all Asians, Hispanics, and Blacks.

The FDA has recently approved ultrasound devices (Sahara and Achilles) for BMD determinations (osteosonometry) at the calcaneus in order to help screen for osteoporosis and to predict fracture risk. Other imaging methods are also available for assessing the skeleton both axially and peripherally. Peripheral measurements may be useful in some elderly patients in whom calcification and other artifacts interfere with DXA. However, peripheral measurements are not useful for screening because they do not measure clinically relevant sites (Table 10.7). Photon beam absorptiometry has been supplanted by these newer methods. In general, there are correlations among all of the methods used to assess the skeleton and bone mass and between different anatomical sites. However, site-specific studies are the best indicators of fracture risk at a particular site. Thus, although BMD at one site correlates with BMD at another site and has some predictive value for fracture risk at another site (Table 10.2), BMD at the hip, spine, or wrist respectively has the best predictive value for fracture at those corresponding sites.

While evidence of a low BMD does encourage women to start [estrogen] therapy, there is no evidence of an effect of subsequent measurements on compliance.

Densitometry studies can be used to monitor the effects of therapy in osteoporosis and other diseases. The precision of the measurement becomes even more important in such circumstances. In research studies, precisions of 0.5% can be achieved. However, in clinical practice, the precision can range up to 2% to 3%. In general, for a follow-up BMD to be significant, the change must exceed 2.8 times the site-specific precision error (%), which is about 0.75% for the spine and 1% to 1.25% for the femur. Thus, for identifying change with a 95% confidence level, a true change in BMD of approximately 3% must occur in the research setting and a change of approximately 6% to 8% must occur in the clinical setting. So repeat studies in osteoporosis should be undertaken after 18 to 24 months. This 2 year figure is consistent with the recently announced (Federal Register, Volume 63, Number 121, 1998) Health Care Financing Administration (HCFA) rules for medicare in which 2 years is the covered interval for repeat measurement. While these rules do allow more frequent monitoring at 3 months for patients with steroid-induced osteoporosis, the general Medicare coverage for BMD according to the Bone Mass Measurement Act of 1998 includes:

- An estrogen-deficient woman at clinical risk for osteoporosis
- An individual with vertebral abnormalities as demonstrated by x-ray to be indicative of osteoporosis, low bone mass (osteopenia), or vertebral fracture
- An individual receiving or expected to receive a glucocorticoid therapy equivalent of 7.5 mg of prednisone, or greater, per day for more than 3 months, or if the expected duration of such therapy is more than 3 months
- An individual with primary hyperparathyroidism (PH)
- An individual being monitored to assess the response to or efficacy of an FDA-approved osteoporosis drug therapy.

Medicare will cover bone mass measurements once every 2 years and more frequently if medically necessary. Covered technologies include all FDA-approved procedures.

■ Bone Markers

Measurement of bone markers is an evolving clinical tool in the evaluation and management of the patient with osteoporosis. The basic biology of bone markers is discussed in Chapter 1, *Skeletal Metabo-*

lism and illustrated in Figure 1.4. Their use, both actual and theoretical, is summarized in Table 1.10.

Bone markers do not yet have the accuracy and precision necessary for diagnostic value. Any clinical value of bone markers in osteoporosis is for management, not diagnosis (Table 1.10). There is substantial physiological variability in bone-marker measurements among different individuals, ranging from 15% to 40% depending on the specific maker, with resorption markers having the greatest variability. Furthermore, the inherent precision of bone-marker testing is not comparable to that of the alternative for skeletal assessment, bone densitometry. The precision of DXA ranges from 0.5% to 4%, with the lower values in research protocols and the higher values in clinical application. By contrast, the precision for tests of most biochemical markers of bone turnover range from 10% to 40%. Such precision is comparable to most biochemical tests, but falls far short of the precision of bone densitometry procedures, especially DXA.

Despite their limitations, especially in diagnosis, bone markers can play a role in decisions about treatment if they have several key characteristics:

- The ability to distinguish between osteoporotic patients and age-matched controls
- Correlation to better established skeletal measurements such as BMD
- The ability to respond to and reflect the efficacy of the different treatments available for osteoporosis.

Many of the bone markers discussed in Chapter 1, *Skeletal Metabolism*, either have not been extensively studied in osteoporosis, can not distinguish between normals and the patient with osteoporosis, and/or have not been shown to reflect changes in bone metabolism subsequent to therapy. Even the best markers are increased only a few-fold in osteoporotics compared to normals. Since current FDA-approved therapies for osteoporosis are antiresorptive agents, these considerations are particularly relevant for markers of bone resorption.

With a careful consideration of their limitations, several bone markers have been applied to clinical studies of osteoporosis, and FDA-approved indications for their clinical application to HRT and bisphosphonate treatment are beginning to appear (Table 1.10). This is especially so for the formation marker, bone-specific alkaline phosphatase (BAP) and the resorption markers, N-terminal cross-linked telopeptide of type I collagen (NTX) and deoxypyridinoline (DPD). Markers can help to identify the "high turnover" patient and to determine compliance and/or response to therapy. The former may be used to make decisions about initiating therapy and the latter about continuing therapy.

Despite the fact that bone markers have less precision than BMD, they can still be useful in reflecting drug therapy, especially antiresorptives. For example, the bisphosphonates can inhibit bone resorption markers by up to 50% and thus demonstrate changes in marker whose variability, although substantial, is less. And although the variability of NTX (and the cross-linked C-terminal telopeptide of type I collagen [CTX]) is greater than DPD, drug-induced suppression of the former (50%) is greater than the latter (30%). So both are of approximately equal value in reflecting antiresorptive therapy.

A decision about initiating therapy in the patient with borderline BMD may be furthered by the finding of abnormal bone markers and vice versa, especially bone resorption markers and antiresorptive therapy. But abnormal bone markers in the absence of reduced BMD are not indications for treatment.

Compliance with therapy and response to therapy may be assessed within weeks by resorption markers and a few months by formation markers, while many months to years are needed to do so for tests of BMD. Compliance with therapy must not be confused with effectiveness of therapy, which is the function of BMD measurement. Finally, bone markers may be useful in arriving at the optimal drug dose and schedules for treatment. This potential application is especially the case for estrogens and CT, where there remains a range of apparently effective regimens.

The predictive value of bone markers for fractures and bone loss is under intense study but not yet firmly established. While there are some studies to indicate that abnormal resorption markers can predict the loss of bone mass as measured by BMD, fracture prediction is tenuous. However, BAP has FDA-endorsement for predicting BMD response to estrogen replacement therapy (ERT). Furthermore, there are some recent studies to suggest that some markers of bone resorption and formation, especially the former, may be able to predict bone fractures. This predictive value may be enhanced by considering bone markers along with BMD measurements. Bone markers may also predict BMD response to therapy. In a study of HRT in postmenopausal females, BMD after 1 year of treatment was maintained or increased in 88% of patients demonstrating at 6 months a decline in NTX of 30% or greater.

While recent results such as these are promising and likely to be more positive with improved markers, the practitioner should recall the tenuous nature of any serial correlations among bone marker changes, BMD changes, and fracture risk. More studies are needed to establish clinical applications and indications for tests to measure bone markers. These studies should include substantial numbers of subjects and be performed by unbiased investigators using blinded, randomized protocols. Fracture risk should be the standard rather than only changes in BMD. At this time, the diagnosis of osteoporosis and the prediction of fracture and bone loss risk are best made by BMD, not bone markers.

Prevention and Treatment

There are two general indications for drug use in osteoporosis: treatment and prevention. Treatment refers to preventing further fractures in osteoporotic patients with existing fractures. Prevention refers to prevention of fractures in patients at risk of developing osteoporotic fragility fractures. Thus, patients with decreased BMD can be evaluated for prevention of any fractures and in patients with existing fractures for the prevention of further fractures.

Hormone replacement therapy with estrogens has been the mainstay of the treatment and/or prevention of osteoporosis and its fractures. However, several newer agents have recently become available. All of the FDA-approved agents are antiresorptive drugs (Table 10.8). Although even newer antiresorptive drugs are on the horizon, recent development efforts have been directed at drugs that stimulate bone formation.

The FDA criterion for approval of new drugs for the treatment of osteoporosis is the ability to reduce fracture rates, but newer estrogen-related drugs have been held to the standard of BMD changes. Furthermore, proof of efficacy in reducing fracture rate must be demonstrated in randomized, clinical trials. Although all fractures are considered, special emphasis is placed on fractures of the hip and spine, especially the latter for study power considerations. Secondary indices, such as effects on BMD and effects on bone

TABLE 10.8 — FDA-APPROVED DRUGS FOR OSTEOPOROSIS

Generic/*Trade* Drug Name	Manufacturer	Available Preparation	Indication
Estrogens			
Premarin	Wyeth-Ayerst	Injection, tablets	Treatment and prevention
Estraderm	Novartis	Transdermal patch	
Ogen	Pharmacia & Upjohn	Tablets	
Estratab	Solvay	Tablets	
Calcitonin			
Miacalcin	Novartis	Injection, nasal spray	Treatment
Calcimar	Rhône-Poulenc Rorer	Injection	
Calcitonin-Salmon	Astra USA	Injection	
Bisphosphonates			
Alendronate (*Fosamax*)	Merck	Tablets	Treatment and prevention
SERMs			
Raloxifene (*Evista*)	Lilly	Tablets	Prevention

Abbreviations: FDA, Food and Drug Administration; SERM, selective estrogen receptor modulator.

10

markers are not as important elements of the drug approval process.

There is a direct relationship between pretreatment BMD and fracture risk, with the risk of fracture approximately doubling with each 1 SD decrease in BMD. And increased bone turnover (also pretreatment), which can be assessed by bone markers, is associated with increased bone loss and may be an independent risk factor for fracture. However, no quantitative relationship has been established between therapy-related changes in BMD and future fractures, nor for bone markers. Thus, some therapies may decrease the risk of future fractures without substantial increases in BMD.

Most agents that inhibit bone resorption in osteoporosis also result in a significant increase in BMD within several months of administration. This is due to the fact that antiresorptive therapy decreases the rate of initiation of new bone remodeling cycles, resulting in fewer remodeling sites and a decrease in remodeling space. It is the filling in of this remodeling space that accounts for the increase in BMD produced by most antiresorptive agents. This process continues for several years, after which time BMD reaches a plateau.

Vertebroplasty has been recently introduced for the symptomatic treatment of vertebral fracture. The technique involves the percutaneous injection of polymethylmethacrylate into the involved vertebral body.

■ Estrogens

Estrogen has been deemed by the FDA as effective for the treatment of postmenopausal osteoporosis since 1986, mostly on the basis of epidemiological studies. Estrogens are now indicated for both the prevention of osteoporosis and treatment of established osteoporosis. HRT indicates the administration of estrogens and progestins. ERT indicates the administration of unopposed estrogens, which is now rarely done because of cancer risks. Estrogens act to inhibit bone resorption, although the mechanism of action is unclear in that osteoclasts have relatively few, if any, estrogen receptors. Therefore, estrogen may mediate its antiresorptive effects by regulating the production by other receptor-positive cells, such as osteoblasts or immune cells, of bone active cytokines, lymphokines, and growth factors. In addition to inhibiting bone resorption, estrogens also act to conserve calcium at the gut and kidney.

Several preparations of estrogens have received approval for osteoporosis treatment, although many others are available. They and their daily doses are:

- Conjugated equine estrogens, 0.625 mg (Premarin)
- 17β-Estradiol transdermal patches, 0.05 mg (Estraderm)
- Piperazine estrone sulfate, 0.75 mg (Ogen)
- Esterified estrogens, 0.3 mg, 0.625 mg, and 2.5 mg (Estratab).

Premarin is the most widely studied and used estrogen in the United States and is usually given at a daily dose of 0.625 mg. Estratab is the lowest oral-dose estrogen approved for the prevention of osteoporosis, effective at 0.3 mg/day; it is associated with fewer side effects and no endometrial hyperplasia. Estrogens act to maintain or increase BMD by approximately 1% to 3%. This effect on BMD may occur even up to 10 years postmenopause, but it is greatest when estrogens are initiated within 5 years after the menopause. Once HRT is stopped, bone loss "catches up" over a comparable time interval. This implies that estrogen therapy should be life-long. The addition of progestins to attenuate endometrial hyperplasia does not impair estrogen's effects on reducing bone loss.

Estrogens decrease the occurrence of fractures. Five years of HRT decreases vertebral fractures by 50% to 80% and other fractures, including hip and wrist, by approximately 25%. Ten or more years of estrogen therapy decreases rates of all fractures by approximately 50% to 75%. However, it should be noted that estrogen fracture data have not been developed through the randomized protocols that the FDA now requires for new osteoporosis drugs, but rather on the basis of epidemiological (case-control and cohort) studies. These reservations about estrogen effects also apply to their nonskeletal actions. In fact, a recent randomized study of HRT failed to demonstrate the cardiovascular benefits observed in most epidemiological studies of estrogen.

There are substantial nonskeletal actions of estrogen, and they must be considered in evaluating the osteoporotic patient, both beneficial and harmful effects. Estrogens seem to have beneficial cardiovascular effects that reduce mortality, and they may help with dementia of the Alzheimer's type. A recent study indicated that estrogen therapy may protect against colorectal cancer. However, estrogens can cause withdrawal bleeding, mastodynia, thromboembolic disease, and uterine and breast cancer. The most ominous effect of estrogens is breast cancer; 5 to 10 years of estrogen treatment may increase breast cancer risk 30% to 55%, with no protection from concurrent progestins. By contrast, any estrogen-related risk of endometrial hyperplasia and cancer can be minimized by concurrent progestin therapy. However, it is prudent to periodically monitor the endometrial hyperplasia in the estrogen-treated patient.

Patient-specific considerations are necessary. Although the cardiovascular benefits of estrogens may generally outweigh the breast and uterine cancer risks in population studies, where cardiovascular deaths outweigh breast cancer deaths by approximately 10-

fold, these considerations may not apply to a patient with high breast cancer risk and low cardiovascular risk.

Despite these beneficial effects of estrogen on the skeleton and on other systems, few women continue to take prescribed estrogens for a substantial period of time. In fact, the half-life of estrogen therapy in the United States approximates only 6 months. Furthermore, no more than one third of potential estrogen users in the United States take the drug, and this fraction is much lower in Asian and Third World countries. The reasons for this are complex and cultural and include an aversion to taking pills, dislike of continuing vaginal bleeding, and concerns about cancer, particularly breast cancer. This aversion seems to disregard the fact that estrogen's cardiovascular benefits, including mortality, may substantially outweigh these side effects. Newer preparations of estrogens and estrogen-like agents may improve patient acceptance. Like HRT, alendronate adherence also decreases over time, with almost 50% nonadherence at 1 year. It should be kept in mind that decreased adherence over time occurs with most drugs that are administered for chronic disease, such as antihypertensive and lipid-lowering drugs.

■ Bisphosphonates

Bisphosphonates are stable analogs of pyrophosphate that are deposited in bone because of their affinity for hydroxyapatite. Once in bone, they inhibit its resorption by interfering with osteoclast function and perhaps by stabilizing bone crystals against dissolution. Their osteoclast effects are mediated by inhibiting osteoclast activity and number, the latter by decreasing recruitment into the osteoclast lineage and by promoting osteoclast apoptosis. Several bisphosphonates have been widely used throughout the world in the management of osteoporosis, especially etidronate (Didronel) and alendronate (Fosamax). Only alendronate has FDA approval for this indication, but etidronate is approved for osteoporosis in over 20 other countries. And other bisphosphonates, are undergoing evaluation for osteoporosis in the United States and elsewhere.

Alendronate is indicated for both the treatment and prevention of osteoporosis. For treatment, alendronate at 10 mg/day for 3 years has been shown to increase BMD in the lumbar spine by over 8% and in the hip by close to 6%. In an extensive 3-year study of nearly 1000 women with osteoporosis but no fractures, alendronate resulted in a 48% reduction of new vertebral fractures and a 21% reduction of non-vertebral fractures. In a study of 2000 women with osteoporosis and at least one existing vertebral fracture, alendronate resulted in a 47% reduction in risk of vertebral fractures, a 90% reduction in risk of multiple vertebral fractures, a 51% reduction in risk of hip fractures, and a 48% reduction in risk of wrist fractures. In a fracture intervention trial, alendronate in a lower dose of 5 mg/day for 2 years with an increase to 10 mg/day for 9 months was shown to decrease the rate of new vertebral fractures by 47%. There were similar decreases in hip and wrist fractures.

Alendronate has been extensively documented to reduce osteoporotic fractures. The current FDA criteria for the use of 10 mg daily of alendronate for osteoporosis treatment are a BMD at least 2 SD below the young normal and/or a history of osteoporotic fracture. For prevention, alendronate at 5 mg/day is indicated in postmenopausal females who are at risk of developing osteoporosis. Alendronate is also being evaluated for male osteoporosis and steroid-induced osteoporosis.

The main side effect of alendronate is esophageal irritation. It occurred in up to 6% to 7% of patients during clinical trials and with an apparently greater incidence in clinical practice, sometimes leading to the discontinuation of the drug. Most of the events have been mild, but severe effects such as erosive esophagitis have been observed. The risk of esophagitis can be minimized by taking the drug with a glass of water only while upright for at least 30 minutes before breakfast. Upper GI disorders such as esophagitis, achalasia, and stricture should be considered contraindications to alendronate treatment.

Since alendronate is deposited into bone and remains there for years, there are some concerns about long-term side effects, although none have yet been reported. However, osteomalacia has been reported with doses of etidronate higher than those now recommended for treatment. When the drug is stopped, there is a resumption of the pretreatment decline in BMD that is related to the dose and duration of treatment.

The GI side effects of bisphosphonates can be bypassed by the use of parenteral formulations. For example, pamidronate (Aredia), best known for its antihypercalcemic effect (see Chapter 3, *Hypercalcemia*), has been shown to increase BMD and decrease fractures when given either continuously by mouth or intermittently at 3-month intervals intravenously (IV). Another bisphosphonate, ibandronate, is being developed for administration by IV injection.

■ Calcitonin

Calcitonin is a potent antiresorptive agent because of its effects on inhibiting osteoclast function (see Chapter 2, *Calcium Metabolism*). There are some data to suggest that CT may also act to promote bone formation. Pharmacological preparations of both synthetic human and salmon calcitonin (HCT and SCT) have been available, but only the latter has been FDA approved for osteoporosis treatment. It is indicated for the treatment of osteoporosis in females greater than 5 years postmenopause with a low BMD and who are not candidates for estrogen treatment. It is curious that SCT is more potent in humans than HCT because of its greater affinity for the HCT receptor.

Salmon calcitonin (Miacalcin, Calcimar) is available in both an injectable (all) and intranasal preparation (Miacalcin). The recommended dose for the injectable preparation is 100 units/day and for the intranasal preparation 200 units/day. However, variations of these regimens have been used. SCT increases vertebral bone density by a few percent after several years of treatment. Although some studies show a decrease of only vertebral fractures in patients receiving SCT, these studies involved a relatively small number of patients; extensions of them are ongoing. There are no conclusive data about the effects of SCT on nonvertebral fractures such as fractures of the hip. A relatively unique effect of SCT is its analgesic effect on bone pain, but analgesia is generally difficult to evaluate.

Salmon calcitonin is one of the safest of the drugs available for the treatment of osteoporosis. There is no long-term accumulation of SCT in bone, and it does not decrease, and may even increase, bone quality. Annoying side effects of the injectable form include nausea, vomiting, and flushing, but they may disappear after continuous treatment. Additional side effects of the nasal preparation include mucosal irritation. Resistance to the action of the drugs may occur after long-term use, presumably as the result of antibody development. SCT is also used to treat Paget's disease (see Chapter 8, *Paget's Disease of Bone* [*Osteitis Deformans*]) and hypercalcemia (see Chapter 3, *Hypercalcemia*).

■ Raloxifene: A Selective Estrogen Receptor Modulator

Raloxifene (Evista) has been recently approved by the FDA for the prevention of osteoporosis. Additional studies are underway to establish its use in the treatment of osteoporosis. Raloxifene represents a novel group of drugs that have both estrogen and anti-estrogen effects and are characterized as selective estrogen receptor modulators (SERMs). Thus, raloxifene selectively activates the estrogen receptor in bone and the cardiovascular system but acts as an estrogen receptor antagonist in breast and uterus.

Several studies have demonstrated a beneficial effect for raloxifene in osteoporosis. In 2-year studies of osteoporotic and nonosteoporotic postmenopausal women, raloxifene at 60 mg/day increased BMD in the lumbar spine and hip by 2% to 3% above placebo. A 2-year interim analysis of a vertebral fracture study demonstrated about a 50% reduction in vertebral fractures as well as a 2% to 3% increase in both spine and hip BMD. Although raloxifene was less effective than estrogens in increasing BMD, its effects on vertebral fractures are comparable. Thus, raloxifene is an alternative to estrogens for the prevention of osteoporosis that does not appear to be associated with an increased risk of breast or endometrial cancer. Raloxifene represents a group of drugs that holds considerable promise for not only osteoporosis but for other estrogen-responsive diseases.

Raloxifene also seems to have many of the beneficial lipid-lowering effects of HRT. It reduces total and low-density lipoprotein cholesterol (LDL-c), fibrinogen, and lipoprotein (a) and (c) without raising triglycerides, but has no effect on HDL-c, whereas HRT increases the latter two. Also in contrast to ERT, raloxifene does not cause mastodynia, endometrial hyperplasia, or vaginal bleeding.

Although raloxifene is chemically unrelated to tamoxifen, they both act as SERMs. Tamoxifen, along with agents like droloxifene and idoxifene, are triphenylethylenes, whereas raloxifene is a benzothiopene derivative. Unlike tamoxifen, raloxifene does not cause endometrial hyperplasia, and available evidence suggests that endometrial stimulation may be a class effect of the former. Furthermore, like some but not all studies of tamoxifen, raloxifene seems to substantially protect against breast cancer. Leg cramps and venous thromboembolic events are other side effects of raloxifene. While the latter is a serious adverse effect that is a contraindication to raloxifene, its occurrence rate is similar in women taking HRT or tamoxifen.

The tissue-specific effects of SERMs like raloxifene are not fully understood. There are several newly appreciated molecular mechanisms through which these drugs may be able to have differential effects on estrogen-responsive tissues. They include:

- Differential binding to estrogen receptor isotypes – alpha and beta
- Different tissue distribution of estrogen receptors, with beta having primarily a bone distribution and alpha primarily in reproductive tissues
- Induction of tissue-specific, unique conformational changes in estrogen receptors
- Unique response elements on estrogen-responsive genes
- Differential transcription effects of receptor-associated coregulatory proteins, including:
  – Adaptors
  – Coactivators
  – Corepressors.

■ Fluorides

Fluorides can promote bone formation through their mitogenic effects on osteoblasts. Sodium fluoride at doses of approximately 70 mg/day resulted in an 8%/year increase in lumbar spine BMD during 4 years of study. However, despite the increase in BMD and an increase in bone-formation markers, there was no substantial effect on fractures. In fact, some studies with such high doses of sodium fluoride show an increased fracture rate. This has caused concern about the quality of the new bone stimulated by fluoride.

In US studies with a lower dose and slow-release preparation providing 25 mg/day of fluoride, there was:

- An increase in BMD at spine and hip
- A decrease in new vertebral fractures
- No effect on appendicular fractures.

Subsequent European studies have both confirmed and failed to confirm these findings. Thus, the effects of sodium fluoride on osteoporotic fractures to date is not conclusive. Furthermore, fluoride, especially at higher doses, has substantial side effects, including gastric irritation and an osteoarthritis-like syndrome. Although it has been applied for, a slow-release form of sodium fluoride (Neosten) has not received FDA approval despite recommendation by an FDA advisory committee. But it remains the only drug that increases bone formation to reach this stage in the approval process.

10

Fluoridation of drinking water has no clearly established relationship to osteoporosis.

■ Calcium

Many studies show that treatment of postmenopausal women with calcium supplements decrease rate of bone loss by a few percent and decrease the rate of bone turnover as assessed biochemically. However, the effects of calcium supplementation on fracture rate have been inconsistent and lacking a sufficient number of randomized trials. Some studies even report that high calcium intake may increase the risk of fracture in women and of prostate cancer in men.

However, in contrast to many epidemiological studies, intervention studies often show a beneficial effect of calcium supplementation on BMD and in some studies a reduction of fracture risk at the spine. Furthermore, a recent meta-analysis did find that 1 g/day of calcium supplementation was associated with a 24% reduction in risk of hip fracture. In general, nevertheless, when a beneficial effect of calcium has been shown, it has been slight relative to drug therapy. Calcium supplementation is not without side effects. Constipation is a common occurrence and there is a remote risk of hypercalciuria and kidney stones.

Even though calcium supplementation by itself may not have a substantial effect on BMD and fracture risk in osteoporosis, calcium can improve the effects of other drugs and may allow their dosage to be reduced. Furthermore, some calcium and vitamin D protocols show a reduction in fracture risk without producing a substantial change in BMD. Thus, most clinical trials of osteoporosis now include calcium supplementation, and many add vitamin D.

Patients at risk for osteoporosis, especially elderly females, should be advised to increase an inadequate calcium intake. The importance of calcium for general nutrition and for supporting skeletal growth in adolescents has been previously discussed (see *Nutrition* under *Etiology and Pathogenesis*). Since it is unlikely that adequate dietary calcium can be achieved through food, calcium supplements should be used. Many forms of calcium supplements are available (Table 5.5). The more soluble salts, like citrate, are better absorbed, but more expensive. Calcium carbonate is a widely used compromise.

Divided doses of calcium provide for efficient total absorption. Calcium given at night may inhibit the nocturnal increase in bone resorption. The bioavailability of calcium in calcium-supplemented foods is uncertain, and they should not be considered as adequate for calcium needs until this is established. Deficiency of other bone minerals such as manganese or magnesium may play some role in pathogenesis and treatment.

■ Vitamin D

While vitamin D is the essential treatment for many forms of osteomalacia (see Chapter 7, *Osteomalacia and Rickets*), most studies of the use of vitamin D metabolites for the treatment of osteoporosis do not demonstrate a substantial beneficial effect on BMD or fracture. However, in certain clinical situations characterized by actual or likely vitamin D deficiency, supplementation can reduce fracture risk. For example, in institutionalized females living in northern latitudes with a low dietary calcium intake, dietary supplementation with both calcium and vitamin D can increase BMD at the femoral neck and reduce the incidence of hip fracture. Other mechanisms by which the aging process causes vitamin D deficiency include: 10

- Decreased sunlight exposure
- Decreased skin stores of precursors
- Decreased renal activation.

It seems prudent, therefore, to assure that all patients who are at risk for osteoporosis have an adequate vitamin D as well as calcium dietary intake. This is best assessed by measuring 25-D in the serum and treating appropriately. Calcitriol and other related metabolites have not been shown to be better than vitamin D for this indication, so standard preparations can be recommended (Table 5.4). Artificial UV radiation (with special lamps) can be used as a supplementary procedure.

■ Other Agents

Several new therapies are under evaluation for the treatment and prevention of osteoporosis (Table 10.9). They include new bisphosphonates, SERMs, and vitamin D metabolites; bone-growth factors and cytokines, growth hormone and its secretagogues, strontium salts, ipriflavone, and PTH.

TABLE 10.9 — NEW DRUGS FOR OSTEOPOROSIS
New Preparations of Existing Therapies • Calcium • Vitamin D • Calcitonin • Estrogen • Bisphosphonates • SERMs
New Experimental Agents • Fluorides, including slow release • PTH 1-34 and related PTH and PTHrP peptides • Strontium • CGRP • Ipriflavone
Abbreviations: SERM, selective estrogen receptor modulator; PTH, parathyroid hormone; PTHrP, parathyroid hormone-related protein; CGRP, calcitonin gene-related peptide.

Parathyroid hormone has received extensive experimental evaluation in animals and recently in humans because of its anabolic effects on bone. Whereas high levels of endogenous PTH are usually associated with increased bone resorption and the bone disease osteitis fibrosa cystica, low and intermittent doses of PTH 1-34, and closely related PTH and parathyroid hormone-related protein (PTHrP) forms, promote bone formation perhaps through an increase in IGF-1 production by osteoblasts. In clinical studies, PTH 1-34 injections increased hip and spine BMD.

Anabolic steroids also promote bone formation, but their use in females is limited by their virilizing effects and in males by their deleterious cardiovascular and prostate effects. However, they may have some use in males and females who are testosterone deficient for their age and gender. Future studies will be needed to define the role of PTH and these other experimental agents in the treatment and/or prevention of osteoporosis.

In addition to calcium and vitamin D, general nutrition should be optimized. Protein supplements may help to attenuate post-fracture morbidity.

■ Alternative Therapies

Several alternative therapies have been touted in the lay press and in the marginal professional press as useful in osteoporosis. These include phytoestrogens, dehydroepiandrosterone (DHEA), and even growth hormone. While growth hormone is beneficial in deficient children and adults, it has no role in growth hormone-sufficient individuals for any indication, including osteoporosis. And there is no credible scientific evidence that phytoestrogens or DHEA have any efficacy in the management of osteoporosis.

Cross-Gender Gonadal Steroids

There is evidence, as discussed earlier, that both estrogens and androgens have some role in maintaining bone mass in both genders. However, there is no evidence that estrogen treatment of males with osteoporosis or androgen treatment of females with osteoporosis is efficacious, and there are substantial side effects and safety concerns.

Other Treatment Protocols and Regimens

Most osteoporotic fractures, especially devastating hip fracture, are caused by a fall, and candidates for osteoporosis are susceptible to falls for many reasons (Table 10.10). Therefore, in addition to drug therapy and dietary supplementation, other more mundane measures should be used to prevent falls and should be instituted. These measures are summarized in Table 10.11.

TABLE 10.10 — FACTORS PREDISPOSING TO FALLS
• Neurological disease • Sensory disorders • Muscle weakness • Medications interfering with balance • Chronic illness

National Osteoporosis Foundation

The NOF has recently issued a guide about cost-effective osteoporosis management. They summarized the effect of different treatments on percentage reduction of the risk of osteoporotic fracture as follows:

- Calcium and vitamin D – 10% reduction (all sites)
- CT – 0% (hip, wrist), 75% (spine)
- HRT – 25% (hip, wrist), 50% (spine)
- Bisphosphonates – 50% (all sites).

In considering osteoporosis risk factors, among them BMD, age, a personal or family history of osteoporotic fracture, smoking, alcohol use, and thinness (lowest quartile), the NOF recommended the following guidelines regarding cost effectiveness in reducing osteoporotic fracture risk:

TABLE 10.11 — PROCEDURES FOR MINIMIZING FALLS AND THEIR IMPACT
Lighting • Minimize glare • Minimize shadows
Flooring • Avoid scatter rugs • Avoid slippery surfaces • Avoid electrical cords
Stairs • Railings • Well-lighted • Nonslippery surfaces
Bathroom • Grab bars • Nonslippery bath and shower • Seat in shower
Kitchen • Appliances well positioned • Cupboards and shelves within reach
Footwear • Nonskid sole • Well fitting shoes
Hip Padding

10

- Cost-effective use of BMD:
  - Women ≥ age 65 who are willing to accept treatment if their BMD is low
  - Women < age 65 with one or more osteoporotic risk factor
  - Women with vertebral or hip fractures in order to confirm the diagnosis of osteoporosis
  - Women on HRT for prolonged periods
- Cost-effective treatment populations of menopausal women with:
  - Vertebral or hip fractures

– Risk factors and a T-score below 1.5
– T-score below 2.0.

The most cost-effective treatment is HRT followed by alendronate. CT can be used when these drugs cannot. The beneficial effects of raloxifene had not been established, as they are now, by the time of this evaluation. It should be kept in mind that cost effectiveness is only one of several ways to assess treatment regimens.

Secondary Osteoporosis

Secondary osteoporosis is the unsatisfying name given to a poorly defined group of diseases characterized by decreased bone mass and increased fracture risk and distinguishable from primary osteoporosis. In this group of diseases, a specific etiological agent can often be found or postulated. The features of the osteoporosis are usually dominated by the effects of the offending agent, and treatment is usually directed at the agent. For example, the osteoporosis associated with thyroid hormone excess is accompanied by the clinical and biochemical manifestations of overt or latent hyperthyroidism, and treatment is directed at the thyroid hormone excess. Secondary causes of osteoporosis account for a relatively small percentage of this group of bone diseases, and these causes are summarized in Table 10.12.

■ Steroid-Induced Osteoporosis

A most important cause of secondary osteoporosis is glucocorticoid excess, usually due to the administration of these drugs for their anti-inflammatory effect in a variety of disease states but also seen in Cushing's syndrome. Among the most common are chronic lung disease and rheumatoid arthritis with organ transplantation becoming increasingly important.

TABLE 10.12 — CAUSES OF SECONDARY OSTEOPOROSIS
Drugs and Medicines • Glucocorticoids and other immunosuppressants • Thyroid hormone • Anticoagulants • Calciuretics
Endocrine and Metabolic Diseases • Hyperthyroidism • Hyperparathyroidism • Hypogonadism • Cushing's syndrome • Diabetes mellitus • Hypopituitarism • Anorexia nervosa • Homocystinuria
Malabsorption
Mastocytosis
Lympho- and Myeloproliferative Disease • Lymphoma • Myeloma
Spinal Cord Injury (immobilization)
Space Flight (weightlessness)
Certain Arthritides (eg, rheumatoid arthritis)

Glucocorticoids have multiple deleterious effects on the skeleton, including:

- Inhibition of bone formation (osteoblast)
- Promotion of osteoblast apoptosis
- Increased bone resorption (osteoclast)
- Decreased GI absorption of calcium
- Increased renal excretion of calcium
- Secondary increase in PTH
- Decreased production of bone growth factors
- Decreased collagen and osteocalcin synthesis
- Decreased gonadal hormone production

- Decreased growth hormone, IGF-1, and TGF-β
- Cause osteonecrosis
- Generally catabolic.

They also stimulate the production of cytokines and lymphokines that increase bone resorption and decrease bone formation.

The severity of the osteoporosis is directly related to the dose and duration of the steroid treatment. In general, treatment with prednisone ≥ 7.5 mg/day or its equivalent for 3 months or more will result in osteoporosis. Trabecular bone is more involved than cortical bone. However, patients may be variable in their response, so they should be continually monitored, especially since steroid-induced bone loss is not fully reversible, even in the rare circumstance where the steroid can be discontinued. No therapeutic dosage of steroid has been identified that does not have deleterious effects on bone. Even inhaled steroids have such effects. Measurement of BMD and bone markers can be useful for monitoring. The changes in these steroid-induced measurements occur within months and are much greater than seen in primary osteoporosis. Resorption markers can increase within weeks of steroid administration.

Steroid-induced osteoporosis can be considered a subset of immunosuppressant-induced osteoporosis in that most other immunosuppressant agents, including cyclosporine, azathioprine, and methotrexate, also produce a similar form of osteoporosis. The major effect of these drugs, like glucocorticoids, is to increase bone resorption. The relative importance of this form of osteoporosis will increase with the increasing success of heart, kidney, and other transplantation procedures that require immunosuppression.

■ Treatment of Steroid-Induced Osteoporosis

The direct treatment of steroid-induced osteoporosis is to decrease the dose of steroids or entirely eliminate them. However, this is often difficult or impossible because of their beneficial effect in controlling the manifestations of the underlying disease. Many regimens have been tried to treat or control steroid-induced osteoporosis, including calcium and vitamin D metabolites, but they have in general not been beneficial. Some studies have indicated that SCT may help in steroid-induced osteoporosis. But most of the promise for effective treatment comes from the more potent bisphosphonates.

Both alendronate and etidronate have been shown to help attenuate steroid-induced osteoporosis in clinical trials. In one recent study of males and females recently begun on corticosteroids, four cycles of etidronate 400 mg for 14 days followed by calcium 500 mg for 10 weeks resulted in an increased spine BMD and a substantial reduction of new vertebral fractures. In another recent study, alendronate was evaluated for the treatment and prevention of glucocorticoid-induced osteoporosis. Over a 10-month period in a randomized, placebo-controlled study in both genders, alendronate resulted in an increased BMD at the spine and hip, a decrease in markers of bone turnover, and a decrease in new vertebral fractures. Beneficial effects of alendronate on BMD in patients with Cushing's syndrome has also been demonstrated. Randomized studies of several newer drugs are underway to evaluate their effect in both the treatment and prevention of steroid-induced osteoporosis.

Since glucocorticoids and other immunosuppressants induce bone loss, in part by increasing bone resorption, it is reasonable to hypothecate that most antiresorptive agents will be helpful in treatment and/or prevention of immunosuppressant osteoporosis. In fact, beneficial results have been presented in small trials for most osteoporosis drugs and regimens, including fluorides. While promising, these studies should be extended by randomized protocols in order to identify effective treatment and prevention for a form of osteoporosis whose incidence and prevalence are likely to increase. In the meantime, all patients on steroids should be managed with the smallest dose of glucocorticoid needed to control the underlying disease. Adequate dietary calcium and vitamin D should be provided. Calcitriol with its risk of hypercalcemia and hypercalciuria can be supplanted for this indication by ordinary vitamin D. A dose of approximately 50,000 IU/week can often raise the serum 25-D to the desired level of about 30 ng/dL. Selected patients can be further treated with available formulations of SCT and bisphosphonates.

Glucocorticoids can suppress gonadal function, especially testosterone production in males. So the appropriate serum measurements should be made in patients on steroid therapy and, if indicated, testosterone or estrogen treatment should be instituted. The preferred measurements are free or bioavailable steroids. Table 10.13 summarizes this approach.

TABLE 10.13 — MANAGEMENT OF STEROID-INDUCED OSTEOPOROSIS

- Minimize the dose of steroids
- Assure adequate calcium and vitamin D intake
- Conserve renal calcium with a thiazide diuretic
- Institute gonadal steroids if suppressed
- Institute antiresorptive therapy:
  - Bisphosphonates
  - Calcitonin (CT)
  - Estrogens and selective estrogen receptor modulators (SERMs)
- Monitor with bone mineral density (BMD) and bone markers
- Minimize other risk factors

Summary

Osteoporosis is the most common skeletal disease, affecting tens of millions of individuals and costing health-care systems billions of dollars annually. The porous skeleton has:

- Decreased mass and density
- Increased fragility
- Fractures with little trauma.

Although it most commonly involves postmenopausal Caucasian females, osteoporosis can affect all races and both genders. The etiology of osteoporosis is unknown, but many risk factors for the primary form of the disease have been identified, including:

- Female gender
- Decreased bone mass
- Increased age
- Decreased lifetime gonadal steroid exposure
- Decreased dietary calcium and vitamin D
- Slight body build
- Family history of osteoporosis
- Personal history of fracture.

All risk factors likely contribute to the development of osteoporosis. Clinical management should focus on modifiable risk factors (Table 10.3).

Less common than primary osteoporosis are secondary forms of the disease that can often be attributed to specific etiologies. Glucocorticoid-induced osteoporosis is the most common and important form of secondary osteoporosis.

Fracture is the most dramatic clinical manifestation of osteoporosis. Although any bone can be involved, the spine, hip and wrist are best appreciated. Osteoporotic fracture is associated with both morbidity and mortality, especially hip fracture. The risk of fracture inversely correlates with bone mass, which is usually quantitated by the measurement by dual energy x-ray absorbtiometry (DXA) of bone mineral density (BMD). Increased bone turnover clinically assessed by bone markers may be an independent risk of fractures. Urinary and serum markers of bone resorption and formation may also help in skeletal assessment, perhaps including fracture risk, and to evaluate management effectiveness and compliance.

The management of primary osteoporosis includes both prevention and treatment of fracture and loss of BMD (Table 10.14). The public health aspects of osteoporosis should be addressed by ensuring that peak bone mass is attained by early adulthood through adequate dietary calcium and vitamin D intake, nutrition, gonadal hormone sufficiency, and exercise. Prevention is directed at those individuals at risk for osteoporosis, especially a low BMD, and treatment is directed at patients with established osteoporosis. Cost effectiveness is one of several criteria for evaluating therapy. It is estimated that the annual cost per life year saved for osteoporosis treatment approximates 30 thousand dollars, a figure comparable for hypertension and lipid-lowering treatments.

Minimizing risk factors is important for both prevention and treatment. All patients should:

- Have adequate intake of calcium
- Have adequate intake of vitamin D
- Participate in regular weight-bearing exercise
- Avoid tobacco use
- Avoid excess alcohol use
- Moderate life-style.

Antiresorptive drugs that are FDA-approved for both prevention and treatment are several estrogens and the bisphosphonate, alendronate. The SERM, raloxifene, is indicated for prevention and salmon calcitonin (SCT) for treatment. Newer forms of these

TABLE 10.14 – SUMMARY OF RELATIVE EFFECTS OF FDA-APPROVED DRUGS IN OSTEOPOROSIS

Agent	BMD Increase		Fracture Risk Reduction	
	Spine	*Hip*	*Spine*	*Hip*
SCT	Yes	N/E	Yes	N/E
HRT/ERT	Yes	Yes	Yes	N/E
Alendronate	Yes	Yes	Yes	Yes
Raloxifene	Yes	Yes	Yes	N/E

Abbreviations: FDA, Food and Drug Administration; BMD, bone mineral density; SCT, salmon calcitonin; HRT, hormone replacement therapy; ERT, estrogen replacement therapy; N/E, not established.

As discussed in the text in more detail, this comparison has many limitations because of differences among the studies. They include different agents, durations, populations, and methods of analysis. Unlike the other agents, there are essentially no placebo-controlled, randomized, prospective, clinical trials of HRT (estrogen, progestin)/ERT (estrogen) therapy. Such SCT studies are only of osteoporosis treatment while the other agents are for treatment and prevention. BMD increase does not necessarily correlate with fracture risk reduction, and one can occur without the other; FDA approval emphasizes fracture risk reduction. Valid comparisons can only be made by studies in which different agents are directly compared in the same protocol. Thus, this table only indicates whether or not an effect is present or has not been established (N/E).

agents and new agents are being evaluated for prevention and treatment of primary osteoporosis.

Hormone replacement therapy (HRT) has been considered the first-line approach to osteoporosis treatment; alendronate and CT are considered next, to be used when HRT cannot be used for the many reasons discussed previously. However, a growing body of new studies is putting HRT, alendronate, and raloxifene on a par for fracture risk reduction. Raloxifene is a representative of a group of important new agents, the SERMs. These agents are of unique utility in osteoporosis management, especially for women who will not or cannot take estrogens or do not need the hot-flush suppressing actions of estrogens. Raloxifene should be considered for the osteoporotic, postmenopausal woman without significant vasomotor symptoms, and especially for the patient in whom breast and uterine cancer are substantial concerns. Combinations of drugs, such as bisphosphonates and estrogens, seem to have additive effects in treatment. Approximate annual costs of regimens are:

- Calcium and vitamin D – $50
- Estrogens – $500
- CT – $750
- Bisphosphonates – $750.

The treatment of secondary osteoporosis is directed at the etiological agent and/or its effects. Bisphosphonates seem to be effective in the treatment and prevention of steroid-induced osteoporosis. In evaluating new drugs for osteoporosis, the physician should regard fracture reduction as the single most important criterion for efficacy. Effects on BMD and bone markers are secondary considerations.

The current emphasis on osteoporosis as one of the most common diseases of the elderly should lead to further improvement in diagnosis and clinical management. National Osteoporosis Foundation (NOF) recommendations can serve as a guide to clinical management.

SUGGESTED READING

Adachi JD, Bensen WG, Brown J, et al. Intermittent etidronate therapy to prevent corticosteroid-induced osteoporosis. *N Engl J Med.* 1997;337:382-387.

Adachi JD, Bensen WG, Bell MJ, et al. Salmon calcitonin nasal spray in the prevention of corticosteroid-induced osteoporosis. *Br J Rheumatol.* 1997;36:255-259.

Abellan Perez M, Bayina Garcia FJ, Calabozo M, et al. Multicenter comparative study of synthetic salmon calcitonin administered nasally in the treatment of established postmenopausal osteoporosis. *An Med Interna.* 1995:12:12-16.

Agnusdei D, Gonnelli S, Camporeale A, et al. Clinical efficacy of treatment with salmon calcitonin, administered intranasally for 1 year, in stabilized postmenopausal osteoporosis. *Minerva Endocrinol.* 1989;14:169-176.

Ahmed AIH, Ilic D, Blake GM, Rymer JM, Fogelman I. Review of 3,530 referrals for bone density measurements of spine and femur: evidence that radiographic osteopenia predicts low bone mass. *Radiology.* 1998;207:619-624.

Anderson FH, Francis RM, Selby PL, Cooper C. Sex hormones and osteoporosis in men. *Calcif Tissue Int.* 1998;62:185-188. Editorial.

Aoyagi K, Ross PD, Davis JW, Wasnich RD, Hayashi T, Takemoto T-I. Falls among community-dwelling elderly in Japan. *J Bone Miner Res.* 1998;13:1468-1474.

Avioli LV. The future of ipriflavone in the management of osteoporotic syndromes. *Calcif Tissue Int.* 1997;61(suppl 1):S33-S35.

Baker VL, Draper M, Paul S, et al. Reproductive endocrine and endometrial effects of raloxifene hydrochloride, a selective estrogen receptor modulator, in women with regular menstrual cycles. *J Clin Endocrinol Metab.* 1998;83:6-13.

Bilezikian JP, Morishima A, Bell J, Grumbach MM. Increased bone mass as a result of estrogen therapy in a man with aromatase deficiency. *N Engl J Med.* 1998;339:599-603.

Black DM, Cummings SR, Karpf DB, et al. Randomised trial of effect of alendronate on risk of fracture in women with existing vertebral fractures. Fracture Intervention Trial Research Group. *Lancet.* 1996;348:1535-1541.

Black DM, Reiss TF, Nevitt MC, Cauley J, Karpf D, Cummings SR. Design of the Fracture Intervention Trial. *Osteoporos Int.* 1993;3(suppl 3):S29-S39.

Brismar TB, Ringertz H. Gradient-echo magnetic resonance signal decay in a porcine vertebral body model: influence of chemical shift. *Acad Radiol.* 1997;4:43-48.

Bryant HU, Dere WH. Selective estrogen receptor modulators: an alternative to hormone replacement therapy. *Proc Soc Exp Biol Med.* 1998;217:45-52.

Burckhardt P, Burnand B. The effect of treatment with calcitonin on vertebral fracture rate in osteoporosis. *Osteoporos Int.* 1993;3:24-30.

Canalis E. Inhibitory actions of glucocorticoids on skeletal growth. Is local insulin-like growth factor I to blame? *Endocrinology.* 1998;139:3041-3042. Editorial.

Carani C, Qin K, Simoni M, et al. Effect of testosterone and estradiol in a man with aromatase deficiency. *N Engl J Med.* 1997;337: 91-95.

Chel VGM, Ooms ME, Popp-Snijders C, et al. Ultraviolet irradiation corrects vitamin D deficiency and suppresses secondary hyperparathyroidism in the elderly. *J Bone Miner Res.* 1998;13:1238-1242.

Cheng XG, Lowet G, Boonen S, Nicholson PHF, van der Perre G, Dequeker J. Prediction of vertebral and femoral strength *in vitro* by bone mineral density measured at different skeletal sites. *J Bone Miner Res*: 1998;13:1439-1443.

Chesnut CH, Bell NH, Clark GS, et al. Hormone replacement therapy in postmenopausal women: urinary N-telopeptide of type I collagen monitors therapeutic effect and predicts response of bone mineral density. *Am J Med.* 1997;102:29-37.

Chiodini I, Carnevale V, Torlontano M, et al. Alterations of bone turnover and bone mass at different skeletal sites due to pure glucocorticoid excess: study in eumenorrheic patients with Cushing's syndrome. *J Clin Endocrinol Metab.* 1998;83:1863-1867.

Cohen-Solal ME, Boitte F, Bernard-Poenaru O, et al. Increased bone resorbing activity of peripheral monocyte culture supernatants in elderly women. *J Clin Endocrinol Metab.* 1998;83:1687-1690.

Colucci S, Minielli V, Zambonin G, et al. Alendronate reduces adhesion of human osteoclast-like cells to bone and bone protein-coated surfaces. *Calcif Tissue Int.* 1998;63:230-235.

Cosman F, Lindsay R. Is parathyroid hormone a therapeutic option for osteoporosis? A review of the clinical evidence. *Calcif Tissue Int.* 1998;62:475-480.

Cosman F, Nieves J, Woelfert L, Gordon S, Shen V, Lindsay R. Parathyroid responsivity in postmenopausal women with osteoporosis during treatment with parathyroid hormone. *J Clin Endocrinol Metab.* 1998;83:788-790.

Cumming RG, Nevitt MC. Calcium for prevention of osteoporotic fractures in postmenopausal women. *J Bone Miner Res.* 1997;12: 1321-1329.

Cummings SR, Browner WS, Bauer D, et al. Endogenous hormones and the risk of hip and vertebral fractures among older women. *N Engl J Med.* 1998;339:733-738.

Cummings SR, Black DM, Nevitt MC, et al. Bone density at various sites for prediction of hip fractures. *Lancet.* 1993;341:72-75.

Cummings SR, Black DM, Vogt TM. Changes in BMD substantially underestimate the anti-fracture effects of alendronate and other antiresorptive drugs. *J Bone Miner Res.* 1996;11(suppl 1):S102.

Cummings SR, Norton L, Eckert S, et al. Raloxifene reduces the risk of breast cancer and may decrease the risk of endometrial cancer in post-menopausal women. Two-year findings from the Multiple Outcomes of Raloxifene Evaluation (MORE) trial. *American Society of Clinical Oncology Program/Proceedings*. Thirty-fourth Annual Meeting. May 16-19, 1998. Los Angeles, CA.

Dawson-Hughes B. Osteoporosis treatment and the calcium requirement. *Am J Clin Nutr.* 1998;67:5-6. Editorial.

Deftos LJ. Bone protein and peptide assays in the diagnosis and management of skeletal disease. *Clin Chem.* 1991;37:1143-1148.

Deftos LJ, Schiff SJ. Predicting PTH pulses and patterns in osteoporosis. *J Clin Invest.* 1995;95:2433-2434.

Delmas PD, Bjarnason NH, Mitlak BH, et al. Effects of raloxifene on bone mineral density, serum cholesterol concentrations, and uterine endometrium in postmenopausal women. *N Engl J Med.* 1997; 337:1641-1647.

Deroisy R, Collette J, Albert A, et al. Comparison of the short-term effects of three oral calcium-vitamin D formulations and placebo on calcium metabolism. *Curr Ther Res.* 1998;59;370-378.

Dimai HP, Porta S, Wirnsberger G, et al. Daily oral magnesium supplementation suppresses bone turnover in young adult males. *J Clin Endocrinol Metab.* 1998;83:2742-2748.

Di Somma C, Colao A, Pivonello R, et al. Effectiveness of chronic treatment with alendronate in the osteoporosis of Cushing disease. *Clin Endocrinol.* 1998;48:655-662.

Eastell R. Treatment of postmenopausal osteoporosis. *N Engl J Med.* 1998;338:736-746.

Ebeling PR, Erbas B, Hopper JL, Wark JD, Rubinfeld AR. Bone mineral density and bone turnover in asthmatics treated with long-term inhaled or oral glucocorticoids. *J Bone Miner Res.* 1998;13: 1283-1289.

Eisman JA. Genetics, calcium intake and osteoporosis. *Proc Nutr Soc.* 1998;57:187-193.

Emery JG, McDonnell P, Burke MB, et al. Osteoprotegerin is a receptor for the cytotoxic ligand trail. *J Biol Chem.* 1998;273: 14363-14367.

Epstein RS, Feng W, Hirsch LJ, Kelly M. Intervention thresholds for the treatment of osteoporosis: comparison of different approaches to decision-making. *Osteoporos Int.* 1998;8(suppl 1):S22-S27.

Ettinger B. Guidelines for established osteoporosis treatment. *Menopause Management.* 1997;6:6-9.

Ettinger B. Overview of estrogen replacement therapy: a historical perspective. *Proc Soc Exp Biol Med.* 1998;217:2-5.

Ettinger B, Black D, Cummings S, et al for the MORE Study Group. Raloxifene reduces the risk of incident vertebral fractures: 24-month interim analyses. *Osteoporos Int.* In press.

Ettinger B, Pressman A, Sklarin P, Bauer DC, Cauley JA, Cummings SR. Associations between low levels of serum estradiol, bone density, and fractures among elderly women: the study of osteoporotic fractures. *J Clin Endocrinol Metab.* 1998;83:2239-2243.

Eyre DR, Koob TJ, Van Ness KP. Quantitation of hydroxypyridinium crosslinks in collagen by high-performance liquid chromatography. *Anal Biochem.* 1984;137:380-388.

Fraser WD, Logue FC, Christie JP, et al. Alteration of the circadian rhythm parathyroid hormone and serum phosphate in women with established postmenopausal osteoporosis. *Osteoporos Int.* 1998;8:121-126.

Frediani B, Allegri A, Bisogno S, Marcolongo R. Effects of combined treatment with calcitriol plus alendronate on bone mass and bone turnover in postmenopausal osteoporosis. Two years of continuous treatment. *Clin Drug Invest.* 1998;15:235-244.

Fritz PC, Ward WE, Atkinson SA, Tenebaum HC. Tamoxifen attenuates the effects or exogenous glucocorticoid on bone formation and growth in piglets. *Endocrinology.* 1998;139:3399-3403.

Garnero P, Gineyts E, Arbault P, Christiansen C, Delmas PD. Different effects of bisphosphonate and estrogen therapy on free and peptide-bound bone cross-links excretion. *J Bone Miner Res.* 1995;10:641-649.

Garnero P, Hausherr E, Chapuy MC, et al. Markers of bone resorption predict hip fracture in elderly women: the EPIDOS Prospective Study. *J Bone Miner Res.* 1996;11:1531-1538.

Garnero P, Shih WJ, Gineyts E, Karpf DB, Delmas PD. Comparison of new biochemical markers of bone turnover in late postmenopausal osteoporotic women in response to alendronate treatment. *J Clin Endocrin Metab.* 1994;79:1693-1700.

Genant HK, Lucas J, Weiss S, et al. Low-dose esterified estrogen therapy: effects on bone, plasma estradiol concentrations, endometrium, and lipid levels. Estratab/Osteoporosis Study Group. *Arch Intern Med.* 1997;157:2609-2615.

Genant HK, Wu CY, van Kuijk C, Nevitt MC. Vertebral fracture assessment using a semiquantitative technique. *J Bone Miner Res.* 1993;8:1137-1148.

Gertz BJ, Clemens J D, Holland SD, Yuan W, Greenspan S. Application of a new serum assay for type I collagen cross-linked N-telopeptides: assessment of diurnal changes in bone turnover with and without alendronate treatment. *Calcif Tissue Int.* 1998;63:102-106.

Greenspan SL, Parker RA, Ferguson L, Rosen HN, Maitland-Ramsey L, Karpf DB. Early changes in biochemical markers of bone turnover predict the long-term response to alendronate therapy in representative elderly women: a randomized clinical trial. *J Bone Miner Res.* 1998;13:1431-1438.

Gregg EW, Cauley JA, Seeley DG, Ensrud KE, Bauer DC for the Study of Osteoporotic Fractures Research Group. Physical activity and osteoporotic fracture risk in older women. *Ann Intern Med.* 1998;129:81-88.

Grodstein F, Martinez ME, Platz EA, et al. Postmenopausal hormone use and risk for colorectal cancer and adenoma. *Ann Intern Med.* 1998;128:705-712.

Grodstein F, Stampfer MJ, Colditz GA, et al. Postmenopausal hormone therapy and mortality. *N Engl J Med.* 1997;336:1769-1775.

Guo CY, Johnson A, Locke TJ, Eastell R. Mechanisms of bone loss after cardiac transplantation. *Bone.* 1998;22:267-271.

Hannon R, Blumsohn A, Naylor K, Eastell R. Response of biochemical markers of bone turnover to hormone replacement therapy: impact of biological variability. *J Bone Miner Res.* 1998; 13:1124-1133.

Herrala J, Puolijoki H, Liippo K, et al. Clodronate is effective in preventing corticosteroid-induced bone loss among asthmatic patients. *Bone.* 1998;22:577-582.

Heshmati HM, Khosla S, Burritt MF, O'Fallon WM, Riggs BL. A defect in renal calcium conservation may contribute to the pathogenesis of postmenopausal osteoporosis. *J Clin Endocrinol Metab.* 1998;83:1916-1920.

Hosking D, Chilvers ED, Christiansen C, et al. Prevention of bone loss with alendronate in postmenopausal women under 60 years of age. *N Engl J Med.* 1998;338:485-492.

Hulley S, Grady D, Bush T, et al for the Heart and Estrogen/progestin Replacement Study (HERS) Research Group. Randomized trial of estrogen plus progestin for secondary prevention of coronary heart disease in postmenopausal women. *JAMA*. 1998;280:605-613.

Jeffcoat MK. Osteoporosis: a possible modifying factor in oral bone loss. *Ann Periodontol*. 1998;3:312-321.

Jensen ME, Evans AJ, Mathis JM, Kallmes DF, Cloft HJ, Dion JE. Percutaneous polymethylmethacrylate vertebroplasty in the treatment of osteoporotic vertebral body compression fractures: technical aspects. *Am J Neuroradiol*. 1997;18:1897-1904.

Johansson C, Black D, Johnell O, Odén A, Mellström D. Bone mineral density is a predictor of survival. *Calcif Tissue Int*. 1998;63:190-196.

Jordan VC, Glusman JE, Ecker S, et al. Incident primary breast cancers are reduced by raloxifene: integrated data from multicenter, double-blind, randomized trials in 12,000 postmenopausal women. *American Society of Clinical Oncology Program/Proceedings*. Thirty-fourth Annual Meeting. May 16-19, 1998. Los Angeles, CA.

Kanis JA, Christiansen C, Lindsay R. *Treatment Strategies in Osteoporosis*: *Proceedings of the Fifth International Symposium organized by the Health Council on Osteoporosis, Paris, France, 17-18 October 1997*. New York, NY: Springer; 1998.

Kanis JA, Johnell O, Gullberg B, et al. Evidence for efficacy of drugs affecting bone metabolism in preventing hip fracture. *BMJ*. 1992;305:1124-1128.

Kanis JA, Melton LJ 3rd, Christiansen C, Johnston CC, Khaltaev N. The diagnosis of osteoporosis. *J Bone Miner Res*. 1994;9:1137-1141.

Khosla S, Melaton J III, Atkinson EJ, O'Fallon WM, Klee GG, Riggs BL. Relationship of serum sex steroid levels and bone turnover markers with bone mineral density in men and women: a key role for bioavailable estrogen. *J Clin Endocrinol Metab*. 1998;83: 2266-2274.

Kleerekoper M, Nelson DA. Peripheral bone densitometry: an old friend revisited. *Trans Am Clin Climatol Assoc*. 1998;109:62-70, 70-72.

Kukita T, Hata K, Kukita A, Iijima T. Laminin, a major basement membrane component of the blood vessel, as a negative regulator of osteoclastogensis. *Calcif Tissue Int*. 1998;63:140-142.

Kulak CA, Bilezikian JP. Osteoporosis: preventive strategies. *Int J Fertil Womens Med.* 1998;43:56-64.

Lehmann R, Wapniarz M, Hofmann B, Pieper B, Haubitz I, Allolio B. Drinking water fluoridation: bone mineral density and hip fracture incidence. *Bone.* 1998;22:273-278.

Liberman UA, Weiss SR, Bröll J, et al. Effect of oral alendronate on bone mineral density and the incidence of fractures in postmenopausal osteoporosis. The Alendronate Phase III Osteoporosis Treatment Study Group. *N Engl J Med.* 1995:333:1437-1443.

Luengo M, Pons F, Martinez de Osaba MJ, Picado C. Prevention of further bone mass loss by nasal calcitonin in patients on long term glucocorticoid therapy for asthma: a two year follow-up study. *Thorax.* 1994;49:1099-1102.

Lufkin EG, Riggs BL. Three-year follow-up on effects of transdermal estrogen. *Ann Intern Med.* 1996;125:77. Letter.

Masi L, Becherini L, Colli E, et al. Polymorphisms of the calcitonin receptor gene are associated with bone mineral density in postmenopausal Italian women. *Biochem Biophys Res Commun.* 1998;248:190-195.

Masi L, Becherini L, Gennari L, et al. Allelic variants of human calcitonin receptor: distribution and association with bone mass in postmenopausal Italian women. *Biochem Biophys Res Commun.* 1998;245:622-626. Abstract.

Masud T, Mulcahy B, Thompson AV. Effects of cyclical etidronate combined with calcitriol versus cyclical etidronate alone on spine and femoral neck bone mineral density in postmenopausal osteoporotic women. *Ann Rheum Dis.* 1998;57:378-379.

Mazess RB. Biochemical markers do not predict bone changes in individual patients in response to estrogen. *Am J Med.* 1998;105:80-81. Letter.

McClung MR. New developments in osteoporosis. *Compr Ther.* 1997;23:798-803.

McClung M, Clemmesen B, Daifotis A, et al. Alendronate prevents postmenopausal bone loss in women without osteoporosis. A double-blind, randomized, controlled trial. Alendronate Osteoporosis Prevention Study Group. *Ann Intern Med.* 1998;128:253-261.

McClung M, Yanover M, Menkes CJ, et al. Aledndronate increases bone mass in patients receiving glucocorticoid therapy for pulmonary diseases. *J Allergy Clin Immunol.* 1998;101:S4.

Melton LJ III, Atkinson EJ, O'Fallon WM, Wahner HW, Riggs BL. Long-term fracture prediction by bone mineral assessed at different skeletal sites. *J Bone Miner Res.* 1993;8:1227-1233.

Melton LJ III, Khosla S, Atkinson EJ, O'Fallon WM, Riggs BL. Relationship of bone turnover to bone density and fractures. *J Bone Miner Res.* 1997;12:1083-1091.

Meunier PJ, Sebert JL, Reginster JY, et al. Fluoride salts are no better at preventing new vertebral fractures than calcium-vitamin D in postmenopausal osteoporosis: the FAVOStudy. *Osteoporos Int.* 1998;8:4-12.

Morishima A, Grumbach MM, Simpson ER, Fisher C, Qin K. Aromatase deficiency in male and female siblings caused by a novel mutation and the physiological role of estrogens. *J Clin Endocrinol Metab.* 1995;80:3689-3698.

Nguyen TV, Sambrook PN, Eisman JA. Bone loss, physical activity, and weight change in elderly women: the Dubbo Osteoporosis Epidemiology Study. *J Bone Miner Res.* 1998;13:1458-1467.

Nieves JW, Komar L, Cosman F, Lindsay R. Calcium potentiates the effect of estrogen and calcitonin on bone mass: review and analysis. *Am J Clin Nutr.* 1998;67:18-24.

Okahashi N, Koide M, Jimi E, Suda T, Nishihara T. Caspases (interleukin-1–converting enzyme family proteases) are involved in the regulation of the survival of osteoclasts. *Bone.* 1998;23:33-41.

Ongphilphadhanakul B, Piaseu N, Chailurkit L, Rajatanavin R. Suppression of bone resorption in early postmenopausal women by intransal salmon calcitonin in relation to dosage and basal bone turnover. *Calcif Tissue Int.* 1998;62:379-382.

Orr-Walker BJ, Evans MC, Ames RW, Clearwater JM, Reid IR. Premature hair graying and bone mineral density. *J Clin Endocrinol Metab.* 1997;82:3580-3583.

Orwoll ES, Oviatt SK, McClung MR, Deftos LJ, Sexton G. The rate of bone mineral loss in normal men and the effects of calcium and cholecalciferol supplementation. *Ann Intern Med.* 1990;112:29-34.

Orwoll ES, Weigel RM, Oviatt SK, McClung MR, Deftos LJ. Calcium and cholecalciferol: effects of small supplements in normal men. *Am J Clin Nutr.* 1988;48:127-130.

Overgaard K, Hanson MA, Jensen SB, Christiansen C. Effect of salcatonin given intranasally on bone mass and fracture rates in established osteoporosis: a dose-response study. *BMJ.* 1992;305: 556-561.

Overgaard K, Riis BJ, Christensen C, Pødenphant J, Johansen JS. Nasal calcitonin for treatment of established osteoporosis. *Clin Endocrinol.* 1989;30:435-442.

Overgaard K, Lindsay R, Christiansen C. Patient responsiveness to calcitonin salmon nasal spray: a subanalysis of a 2-year study. *Clin Ther.* 1995;17:680-685.

Pacifici R. Cytokines, estrogen, and postmenopausal osteoporosis – the second decade. *Endocrinology.* 1998;139:2659-2661. Editorial.

Pak CY, Sakhaee K, Adams-Huet B, Piziak V, Peterson RD, Poindexter JR. Treatment of postmenopausal osteoporosis with slow-release sodium fluoride. Final report of a randomized controlled trial. *Ann Intern Med.* 1995;123:401-408.

Peichl P, Griesmacher A, Pointinger P, et al. Association between female sex hormone and biochemical markers of bone turnover in peri-and postmenopausal women. *Calcif Tissue Int.* 1998;62:388-394.

Quesada JM, Alonso J, Gonzalez J, et al. Serum β-2 microglobulin is a marker of high bone remodelling in elderly women. *Mech Ageing Dev.* 1998;102:293-298.

Ravn P, Christensen JO, Baumann M, Clemmesen B. Changes in biochemical markers and bone mass after withdrawal of ibandronate treatment: prediction of bone mass changes during treatment. *Bone.* 1998;22:559-564.

Reddi K, Meghji S, Nair SP, et al. The *escherichia coli* chaperonin 60 (groEL) is a potent stimulator of osteoclast formation. *J Bone Miner Res.* 1998;13:1260-1266.

Reginster JY, Meurmans L, Zegels B, et al. The effect of sodium monofluorophosphate plus calcium on vertebral fracture rate in postmenopausal women with moderate osteoporosis. *Ann Intern Med.* 1998;129:1-8.

Reid IR. Glucocorticoid-induced osteoporosis. *J Clin Densitometry.* 1998;1:65-73.

Ricci TA, Chowdhury HA, Heymsfield SB, Stahl T, Pierson RN Jr., Shapses SA. Calcium supplementation suppresses bone turnover during weight reduction in postmenopausal women. *J Bone Miner Res.* 1998;13:1045-1050.

Rico H, Hernandez ER, Revilla M, Gómez-Castresana F. Salmon calcitonin reduces vertebral fracture rate in postmenopausal crush fracture syndrome. *Bone Miner.* 1992;16:131-138.

Rico H, Revilla M, Hernández ER, Villa LF, Alvarez de Buergo M. Total and regional bone mineral content and fracture rate in postmenopausal osteoporosis treated with salmon calcitonin: a prospective study. *Calcif Tiss Int.* 1995;56:181-185.

Rico H, Revilla M, Villa LF, Hernandez ER, Fernandez JP. Crush fracture syndrome in senile osteoporosis: a nutritional consequence? *J Bone Miner Res.* 1992;7:317-319.

Riggs BL. Vitamin D-receptor genotypes and bone density. *N Engl J Med.* 1997;337:125-126. Editorial.

Riggs BL, Khosla S, Melton LJ III. A unitary model for involutional osteoporosis: estrogen deficiency causes both type I and type II osteoporosis in postmenopausal women and contributes to bone loss in aging men. *J Bone Miner Res.* 1998;13:763-773.

Riggs B, Melton LJ III. The prevention and treatment of osteoporosis. *N Engl J Med.* 1992;327:620-627.

Riggs BL, O'Fallon WM, Muhs J, O'Connor MK, Kumar R, Melton LJ III. Long-term effects of calcium supplementation on serum parathyroid hormone level, bone turnover, and bone loss in elderly women. *J Bone Miner Res.* 1998;13:168-174.

Ringe JD. Treatment of primary osteoporosis with calcium and salmon calcitonin. *Dtsch Med Wochenschr.* 1990;115:1176-1182.

Risteli J, Elomaa I, Niemi S, Novamo A, Risteli L. Radioimmunoassay for the pyridinoline cross-linked carboxy-terminal telopeptide of type I collagen: a new serum marker of bone collagen degradation. *Clin Chem.* 1993;39:635-640.

Rosen CJ. A tale of two worlds in prescribing etidronate for osteoporosis. *Lancet.* 1997;350:1340.

Rosen CJ, Chestnut CH III, Mallinak NJS. The predictive value of biochemical markers of bone turnover for bone mineral density in early postmenopausal women treated with hormone replacement or calcium supplementation. *J Clin Endocrinol Metab.* 1997;82: 1904-1910.

Rosen HN, Moses AC, Garber J, et al. Randomized trial of pamidronate in patients with thyroid cancer: bone density is not reduced by suppressive doses of thyroxine, but is increased by cyclic intravenous pamidronate. *J Clin Endocrinol Metab.* 1998;83: 2324-2330.

Saag KG, Emkey R, Schnitzer TJ, et al. Alendronate for the prevention and treatment of glucocorticoid-induced osteoporosis. *N Engl J Med.* 1998;339:292-299.

Sato Y, Kuno H, Kaji M, Ohshima Y, Asoh T, Oizumi K. Increased bone resorption during the first year after stroke. *Stroke.* 1998;29;1373-1377.

Sato Y, Fujimatsu Y, Kikuyama M, Kaji M, Oizumic K. Influence of immobilization on bone mass and bone metabolism in hemiplegic elderly patients with a long-standing stroke. *J Neurol Sci.* 1998;156:205-210.

Schürch MA, Rizzoli R, Slosman D, Vadas L, Vergnaud P, Bonjour J-P. Protein supplements increase serum insulin-like growth factor-I levels and attenutate proximal femur bone loss in patients with recent hip fracture. A randomized, double-blind, placebo-controlled trial. *Ann Int Med.* 1998;128:801-809.

Simpson E, Davis S. Why do the clinical sequelae of estrogen deficiency affect women more than men? *J Clin Endocrinol Metab.* 1998;83:2214. Letter.

Smith EP, Boyd J, Frank GR, et al. Estrogen resistance caused by a mutation in the estrogen-receptor gene in a man. *N Engl J Med.* 1994:331:1056-1061. Published erratum appears in *N Engl J Med.* 1995;332:131.

Sowers M, Randolph JF Jr, Crutchfield M, et al. Urinary ovarian and gonadotropin hormone levels in premenopausal women with low bone mass. *J Bone Miner Res.* 1998;13:1191-1202.

Stone K, Bauer DC, Black DM, Sklarin P, Ensrud KE, Cummings SR, for the Study of Osteoporotic Fractures Research Group. Hormonal predictors of bone loss in elderly women: a prosepctive study. *J Bone Miner Res.* 1998;13:1167-1174.

Sugimoto H, Kimura T, Oshawa T. Susceptbility effects of bone trabeculae: quantification in vivo using an asymmetric spin echo technique. *Invest Radiol.* 1993;28:208-213.

Szollar SM, Martin EM, Parthemore JG, Sartoris DJ, Deftos LJ. Densitometric patterns of spinal cord injury associated bone loss. *Spinal Cord.* 1997;35:374-382.

Szollar SM, Martin EM, Sartoris DJ, Parthemore JG, Deftos LJ. Bone mineral density and indexes of bone metabolism in spinal cord injury. *Am J Phys Med Rehabil.* 1998;77:28-35.

Tham DM, Gardner CD, Haskell WL. Clinical review 97: potential health benefits of dietary phytoestrogens: a review of the clinical, epidemiological, and mechanistic evidence. *J Clin Endocrinol Metab.* 1998;83:2223-2235.

The Health Council on Osteoporosis. Treatment strategies in osteoporosis. Proceedings of the 5th international symposium. Paris, France. *Osteoporos Int.* 1998;8(suppl 1):S1-S74.

Thomas MK, Lloyd-Jones DM, Thadhani RI, et al. Hypovitaminosis D in medical inpatients. *N Engl J Med.* 1998;338:777-783.

Tilyard MW, Spears GFS, Thomson J, Dovey S. Treatment of postmenopausal osteoporosis with calcitriol or calcium. *N Engl J Med.* 1992;326:357-362.

Treatment strategies in osteoporosis. Fifth International Symposium organized by The Health Council on Osteoporosis. Paris, October 17-18, 1997. *Osteoporos Int.* 1998;8(suppl 1):S1-S74.

Tsushima N, Harada H, Katsumata T, et al. SM-16896, a new bisphosphonate-estrogen hybrid compound, has high affinity to the bone and prevents bone loss in ovariectomized rats. *J Bone Miner Res.* 1997;12:S144.

Walsh BW, Kuller LH, Wild RA, et al. Effects of raloxifene on serum lipids and coagulation factors in healthy postmenopausal women. *JAMA.* 1998;279:1445-1451.

Wehrli FW, Ford JC Attie M, Kressel HY, Kaplan FS. Trabecular structure: preliminary application of MR interferometry. *Radiology*. 1991;179:615-621.

Weinstein RS, Jilka RL, Parfitt AM, Manolagas SC. Inhibition of osteoblastogenesis and promotion of apoptosis of osteoblasts and osteocytes by glucocorticoids. *J Clin Invest*. 1998;102:274-282.

PART IV

OTHER DISEASES OF SKELETAL AND CALCIUM METABOLISM

11

Hypercalciuria and Nephrolithiasis

Introduction

Hypercalciuria is generally defined as a 24-hour urinary calcium greater than 300 mg in males and 250 mg in females. These figures, however, are based on populational studies, and the individual patient with nephrolithiasis may vary from these limits. Hypercalciuria can occur as a consequence of any of the causes of hypercalcemia discussed in Chapter 3, *Hypercalcemia.* Furthermore, the absence of parathyroid hormone (PTH) in certain types of hypoparathyroidism (see Chapter 5, *Hypocalcemia and Hypoparathyroidism*) is permissive for the development of hypercalciuria during treatment with calcium and vitamin D since the hormone acts to decrease renal calcium clearance. This chapter, however, focuses on so-called idiopathic hypercalciuria (IH), a group of disorders in which hypercalciuria occurs in the absence of hypercalcemia and of primary abnormalities of PTH secretion. The formation of kidney stones is the most serious complication of hypercalciuria. IH accounts for a substantial percentage of patients with nephrolithiasis, while primary hyperparathyroidism (PH) and renal tubular disorders account for only a small minority.

Etiology and Pathogenesis

There are three main mechanisms for the development of hypercalciuria (Table 11.1), and they are reminiscent of the mechanisms for the development

TABLE 11.1 — PATHOGENESIS OF HYPERCALCIURIA
Absorptive: *Increased Gastrointestinal Calcium Absorption* • Vitamin D-dependent • Vitamin D-independent
Nonreabsorptive: *Decreased Renal Tubular Reabsorption of Filtered Calcium* • Calcium transport defect • Parathyroid hormone (PTH) • Calcitonin (CT) • Renal tubular acidosis (RTA)
Resorptive: *Increased Bone Resorption* • Primary hyperparathyroidism (PH) • Other causes of increased bone resorption (see Chapter 3, *Hypercalcemia*)

of hypercalcemia. Hypercalciuria usually occurs when there is increased gastrointestinal (GI) absorption of up to 50% of dietary calcium, decreased renal tubular reabsorption of calcium, and/or both. The bone-resorptive form of hypercalciuria results from the associated hypercalcemia and will not be emphasized here. Although stones are due to urine supersaturation with respect to stone crystals, it is important to note that factors other than urinary calcium account for stones; factors such as urinary:

- pH
- Proteins (eg, calgranulin)
- Volume.

The increased calcium absorption in patients with IH is due to increased serum concentrations of 1,25-dihydroxyvitamin D (1,25-D) in many, but not all patients. The cause for increased 1,25-D production is unclear. In some patients, it may be due to an increased PTH secretion that compensates for a renal calcium and/or phosphate leak. However, while many patients with IH do have a low serum phosphorus, few have increased serum PTH.

The defect in tubular reabsorption of calcium that occurs in some patients with IH can be familial and perhaps hereditary. Renal tubular acidosis (RTA) can be characterized by hypercalciuria, but this disease is relatively rare. Calcitonin (CT) can promote the renal excretion of calcium and has been reported as increased in patients with IH. Rather than a primary defect, however, increased CT secretion is more likely to result as a secondary disorder caused by the abnormal postprandial increases in serum calcium seen in the absorptive form of IH.

Signs and Symptoms

Nephrolithiasis and the associated renal colic and its complications are the dominant features of IH. Most stones pass spontaneously. Patients should be instructed to strain their urine to recover the stone for analysis. While kidney stones are usually calcium-containing, such stones can occur in the absence of hypercalciuria. Furthermore, nephrolithiasis can be characterized by noncalcium stones (Table 11.2). Therefore, the patient with nephrolithiasis should also be evaluated for the presence of noncalcium-containing stones and the metabolic diseases that underlie some of them, such as hyperoxaluria and cystinuria. The hypercalciuria can result in bone loss.

Treatment

Since the pathophysiology of the hypercalciuria in the individual patient may not be identified, adequate hydration and perhaps even over hydration is a mainstay for treatment (Table 11.3). Hydration

TABLE 11.2 — COMPOSITION OF RENAL STONES
• Calcium containing:
– Oxalate
– Phosphate
– Mixed
• Noncalcium:
– Struvite (infection)
– Uric acid
– Xanthine
– Dihydroxyadenine
– Cystine
– Drug crystals

should be especially vigorous when there is risk of fluid depletion, such as high summer temperatures and vigorous exercise. There are many anecdotal reports of kidney stones presenting after tennis.

A low calcium diet may not be useful in absorptive hypercalciuria and nephrolithiasis. In fact, dietary calcium restriction may create a negative calcium balance and contribute to a low bone density; it can also worsen oxaluria. So it is prudent to maintain calcium intake at about 800 mg/day. Calcium binders such as cellulose phosphate have been recommended, but they may increase urinary oxalate excretion. Diuretics that augment distal tubular calcium reabsorption can be used when conservative treatment fails (Table 11.3). Certain preparations of phosphorus have been recommended, especially in patients who are hypophosphatemic. Urological procedures such as surgical removal of the stone or lithotripsy can be used for established disease.

Summary

Hypercalciuria can occur because of or independent of hypercalcemia. When independent, it is referred to as idiopathic hypercalciuria (IH). IH can be

TABLE 11.3 — TREATMENT OF HYPERCALCIURIA
• Fluids • Analgesics • Calcium binders – phosphate • Phosphate administration • Calciuretics – Thiazides – Chlorthalidone – Indopimide • Urological procedures

caused by increased gastrointestinal (GI) absorption of calcium, increased renal excretion of calcium, or both. Nephrolithiasis is the clinical consequence of hypercalciuria, but stones can occur in its absence. Treatment with a moderate calcium intake is directed at the absorptive mechanism of IH, and treatment with hydration and calciuretics is directed at the reabsorptive defect. Underlying metabolic causes of nephrolithiasis should be identified. Surgical intervention may be necessary to remove stones.

SUGGESTED READING

Breslau NA, Padalino P, Kok DJ, Kim YG, Pak CY. Physicochemical effects of a new slow-release potassium phosphate preparation (UroPhos-K) in absorptive hypercalciuria. *J Bone Miner Res.* 1995; 10:394-400.

Coe FL, Parks JH. New insights into the pathophysiology and treatment of nephrolithiasis: new research venues. *J Bone Miner Res.* 1997;12:522-533.

Coe FL, Parks JH, Asplin JR. The pathogenesis and treatment of kidney stones. *N Engl J Med.* 1992;327:1141-1152.

Coe FL, Parks JH, Favus MJ. Diet and calcium: the end of an era? *Ann Intern Med.* 1997;126:553-555. Editorial.

D'Angelo A, Calo L, Cantaro S, Giannini S. Calciotropic hormones and nephrolithiasis. *Miner Electrolyte Metab.* 1997;23:269-272.

De Swart PM, Sokole EB, Wilmink JM. The interrelationship of calcium and magnesium absorption in idiopathic hypercalciuria and renal calcium stone disease. *J Urol.* 1998;159:669-672.

Gunther W, Luchow A, Cluzeaud F, Vandewalle A, Jentsch TJ. CIC-5, the chloride channel mutated in Dent's disease, colocalizes with the proton pump in endocytotically active kidney cells. Proceedings of the National Academy of Sciences of the United States of America, 1998 Jul 7, 95(14):8075-8080.

Heilberg IP, Martini LA, Teixeira SH, et al. Effect of etiodronate treatment on bone mass of male nephrolithiasis patients with idiopathic hypercalciuria and osteopenia. *Nephron.* 1998;79:430-437.

Heller HJ, Reza-Albarran AA, Breslau NA, Pak CYC. Sustained reduction in urinary calcium during long-term treatment with slow release neutral potassium phosphate in absorptive hypercalciuria. *J Urol.* 1998;159:1451-1456.

Ivey JL, Roos BA, Shen FH, Baylink DJ. Increased immunoreactive calcitonin in idiopathic hypercalciuria. *Metab Bone Dis Relat Res.* 1981;3:39-42.

Lau AH, Kuk JM, Franson KL. Phosphate-binding capacities of calcium and aluminum formulations. *Int J Artif Organs.* 1998;21:19-21.

Pak CYC. Kidney stones. *Lancet.* 1998;351:1797-1801.

Ruml LA, Dubois SK, Roberts ML, Pak CY. Prevention of hypercalciuria and stone-forming propensity during prolonged bedrest by alendronate. *J Bone Miner Res.* 1995;10:655-662.

Van Faasen A, Van Der Ploeg EMC, Habets HML, Van Der Meer R, Hermus RJJ, Janknegt RA. The effects of the calcium-restricted diet of urolithiasis patients with absorptive hypercalciuria type II on risk factors for kidney stones and osteopenia. *Urol Res.* 1998;26:65-69.

Wasserstein AG. Nephrolithiasis: acute management and prevention. *Disease-A-Month.*1998;44:196-213.

Whiting SJ, Green TJ, MacKenzie EP, Weeks SJ. Effects of excess protein, sodium and potassium on acute and chronic urinary calcium excretion in young women. *Nutr Res*. 1998;18:475-487.

Zerwekh JE, Hughes MR, Reed BY, et al. Evidence for normal vitamin D receptor messenger ribonucleic acid and genotype in absorptive hypercalciuria. *J Clin Endocrinol Metab*. 1995;80:2960-2965.

11

12 Disorders of Magnesium and Phosphorus Metabolism

Introduction

The major disorders of magnesium and phosphorus metabolism, hypomagnesemia with hypocalcemia and hyperphosphatemia with renal failure, have been respectively discussed in Chapters 5, *Hypocalcemia and Hypoparathyroidism* and 9, *Renal Osteodystrophy*. This chapter is a survey of other less common abnormalities of these two minerals.

Etiology and Pathogenesis

The mechanisms that underlie abnormalities in serum magnesium and phosphorus are similar to the mechanisms that underlie abnormalities of serum calcium. Like calcium, both can be abnormally increased or decreased as a consequence of abnormalities of renal excretion and gastrointestinal (GI) absorption. A third regulatory abnormality can be termed redistribution. It refers to the shift of phosphorus and magnesium to and from the intracellular space and is akin to the shift of calcium to and from bone during bone formation and bone resorption that underlies serum calcium disorders. Since there are substantial nonskeletal stores for magnesium and phosphorus, the redistribution abnormalities involve organ systems in addition to bone, especially for phosphorus. The mechanisms of hypophosphatemia and their underlying causes are summarized in Table 12.1 and for hyperphosphatemia in Table 12.2. However, other organ systems may also be involved, especially for

TABLE 12.1 — PATHOGENESIS OF HYPOPHOSPHATEMIA
Decreased Gastrointestinal (GI) Absorption • Vitamin D deficiency • Phosphate-binding antacids • GI disease – malabsorption • Alcohol excess • Dietary deficiency
Decreased Renal Tubular Reabsorption • Primary tubular disorders • Osteomalacia/hypophosphatemia (Table 7.1) • Hyperparathyroidism
Cellular Redistribution • Respiratory alkalosis • Osteoblastic metastases • Hungry bones: postparathyroidectomy • Glucose/insulin

TABLE 12.2 — PATHOGENESIS OF HYPERPHOSPHATEMIA
Decreased Renal Excretion • Renal failure • Increased tubular reabsorption: – Hypoparathyroidism – Tumoral calcinosis – Bisphosphonates – Heparin
Increased Intake • Dietary excess • Vitamin D treatment • Phosphate administration (parenteral)
Cellular Redistribution • Systematic acidosis • Cellular release: – Cancer – Hemolysis – Rhabdomyolysis – Cytotoxic therapy

hypomagnesemia (Table 12.3). A familial form of hypomagnesium has been described.

TABLE 12.3 — PATHOGENESIS OF HYPOMAGNESEMIA
Decreased Gastrointestinal (GI) Absorption • Dietary deficiency • GI disease – malabsorption • Alcohol excess • Vomiting/diarrhea
Increased Renal Excretion • Primary tubular disorder: – Bartter's syndrome – Renal tubular acidosis (RTA) – Gitelman's syndrome • Acquired tubular disorders: – Diuretics – Hypoparathyroidism – Hypercalcemia – Alcohol
Cellular Redistribution • Respiratory alkalosis • Hungry bones/osteoblastic metastasis
Familial Hypomagnesemia

Tables 12.3 and 12.4 respectively summarize the mechanisms for hypomagnesemia and hypermagnesemia. Since magnesium is primarily an intracellular ion, serum measurements may not accurately reflect the magnesium status of the individual. Special tests can be used for this purpose (eg, red blood cell magnesium), but clinical judgment is often necessary to uncover abnormalities in magnesium metabolism (see further discussion in Chapter 5, *Hypocalcemia and Hypoparathyroidism*).

TABLE 12.4 — PATHOGENESIS OF HYPERMAGNESEMIA
Increased Intake • Magnesium-containing antacids • Magnesium-containing laxatives • Magnesium administration (parenteral)
Decreased Renal Excretion • Renal failure • Volume depletion • Certain diuretics • Familial hypocalciuric hypercalcemia (FHH)
Cellular Redistribution • Respiratory acidosis • Metabolic acidosis • Cell lysis

Signs and Symptoms

The signs and symptoms of phosphorus abnormalities are generally those of the underlying disease. For hyperphosphatemia, the underlying disease dominates the clinical picture (Table 12.2). Hypocalcemia and extraskeletal calcification can also result from increased serum phosphorus. Severe hypophosphatemia can be associated with specific clinical findings (Table 12.5).

The signs and symptoms of abnormalities of magnesium metabolism are similar to the signs and symptoms of corresponding abnormalities of calcium metabolisms described in Chapters 3 through 5 (*Hypercalcemia*, *Primary Hyperparathyroidism*, and *Hypocalcemia and Hypoparathyroidism*). In brief, hypomagnesemia is associated with neuromuscular irritability and hypermagnesemia with neuromuscular depression.

TABLE 12.5 — SIGNS AND SYMPTOMS OF HYPOPHOSPHATEMIA
Cardiopulmonary • Impaired cardiac function • Respiratory muscle weakness
Neurologic • Irritability and confusion • Seizures and coma
Musculoskeletal • Muscle weakness, rhabdomyolysis • Osteomalacia
Hematologic • Impaired function of erythrocytes • Impaired function of platelets and leukocytes
Endocrine/Metabolic • Insulin resistance • Increased serum calcium • Decreased parathyroid hormone (PTH), increased 1,25-dihydroxyvitamin D (1,25-D)
Renal • Decreased glomerular filtration rate (GFR) • Hypercalciuria, hypermagnesemia

Treatment

Treatment of hyperphosphatemia is directed against the underlying abnormalities as summarized in Table 12.6. Treatment of hypophosphatemia is directed at replacement, usually accomplished by one of several oral preparations available (Table 12.7). Although corresponding parenteral preparations of phosphorus are available, they can cause extraskeletal calcification and are usually not indicated.

Treatment of hypomagnesemia is described in Chapter 5, *Hypocalcemia and Hypoparathyroidism*. Treatment of hypermagnesemia is directed against re-

TABLE 12.6 — TREATMENT OF HYPERPHOSPHATEMIA
Acute Hyperphosphatemia • Increased renal excretion of phosphorus: – Volume expansion – Bicarbonate – Carbonic anhydrase inhibitors • Shift phosphorus into intracellular space: – Glucose administration – Insulin administration
Chronic Hyperphosphatemia • Dietary phosphorus restriction • Phosphate binders
Dialysis

moving the source. The administration of calcium can provide short-term improvement of clinical status. Dialysis can be useful, especially in the patient with renal failure.

Summary

There are many analogies between the abnormalities of calcium metabolism discussed in Chapters 3 through 5 (*Hypercalcemia*, *Primary Hyperparathyroidism*, and *Hypocalcemia and Hypoparathyroidism*) and the corresponding abnormalities in magnesium and phosphorus metabolism discussed here. Abnormal serum increases in all three are due to essentially the same mechanisms:

- Increases or decreases in renal excretion
- Increases or decreases in gastrointestinal (GI) absorption
- Tissue redistribution.

The latter mechanism refers to bone resorption and bone formation for calcium. However, since magne-

TABLE 12.7 — PHOSPHORUS-CONTAINING ORAL PREPARATIONS

Drug	Manufacturer	Preparation	Phosphorus	Sodium	Potassium
K-Phos	Beach	Tablets	114 mg/tab	—	3.7 mEq/tab
K-Phos Neutral	Beach	Tablets	250 mg/tab	13 mEq/tab	1.1 mEq/tab
Fleet Phospho-Soda	Fleet	Soda	149 mg/mL	24 mEq/tsp (5 mL)	—
Neutra-Phos	Alza	Powder concentrate	250 mg/dose	7.125 mEq/dose	7.125 mEq/dose
		Unit-dose packet	250 mg/dose	7.125 mEq/dose	7.125 mEq/dose
Neutra-Phos-K	Alza	Powder concentrate	250 mg/dose	—	14.25 mEq/dose
		Unit-dose packet	250 mg/dose	—	14.25 mEq/dose

sium and phosphorus have substantial nonskeletal compartments, other tissues can also be involved in their redistribution.

The signs and symptoms of magnesium abnormalities are similar to the corresponding calcium abnormalities (Chapters 3 through 5). Hyperphosphatemia is dominated by the clinical manifestation of the underlying disease, but hypophosphatemia has its own clinical complex (Table 12.5). Treatment is directed at replacement or removal of the specific abnormality.

SUGGESTED READING

Abbott L, Nadler J, Rude RK. Magnesium deficiency in alcoholism: possible contribution to osteoporosis and cardiovascular disease in alcoholics. *Alcohol Clin Exp Res*. 1994;18:1076-1082.

Elisaf M, Bairaktari E, Kalaitzidis R, Siamopoulos KC. Hypomagnesemia in alcoholic patients. *Alcohol Clin Exp Res*. 1998; 22:134.

Levi M, Cronin RE, Knochel JP. Disorders of phosphate and magnesium metabolism. In: Coe FL, Favus MJ, eds. *Disorders of Bone and Mineral Metabolism*. New York, NY: Raven Press; 1992:587-610.

Knochel JP. Mechanisms of rhabdomyolysis. *Curr Opin Rheumatol*. 1993;5:725-731.

Nadler JL, Rude RK. Disorders of magnesium metabolism. *Endocrinol Metab Clin North Am*. 1995;24:623-641.

Rude RK. Magnesium deficiency: a cause of heterogenous disease in humans. *J Bone Miner Res*. 1998;13:749-758.

Shalev H, Phillip M, Galil A, Carmi R, Landau D. Clinical presentation and outcome in primary familial hypomagnesaemia. *Arch Dis Child*. 1998;78:127-130.

Shils ME, Rude RK. Deliberations and evaluations of the approaches, endpoints and paradigms for magnesium dietary recommendations. *J Nutr*. 1996;126(suppl 9):2398S-2403S.

13 Rare Bone Diseases

Introduction

There are several metabolic bone diseases that have importance beyond their rare incidence because of the insights they provide about normal and abnormal skeletal function. Several of them, namely hypophosphatasia and inherited osteomalacias, have been already described in Chapter 7, *Osteomalacia and Rickets*. Several others are briefly discussed here.

Osteopetrosis

Osteopetrosis, stone-like bones, is a descriptive grouping of diseases characterized by petrotic bones caused by failure of osteoclast-mediated bone resorption. There are several forms of the disease, and one arbitrary classification is presented in Table 13.1. All inherited forms are autosomal, some are dominant, most are recessive. They have different presentations and clinical courses, and all are generally characterized by radiodense bones that can be susceptible to fracture. The petrotic bones can impinge on nerves, blood vessels, and marrow space and inhibit their respective functions. The impaired bone resorption is attributed to impaired osteoclast function and can also result in hypocalcemia. The clinical manifestations are based on age of onset and disease progression (Table 13.1). Thus, the disease can be rapidly fatal or have an indolent course.

Although some patients (and some animal models) with the disease have decreased osteoclast num-

TABLE 13.1 — TYPES OF HUMAN OSTEOPETROSIS
• Infantile: – Malignant – Transient – Intermediate • Adult: – Severe – Mild • Carbonic anhydrase type II deficiency

bers, normal or even increased osteoclasts accompanied by decreased function are more common. Involved osteoclasts often have abnormal morphology but can secrete increased amounts of acid phosphatase. This is reminiscent of osteoblasts and alkaline phosphatase (AP) in osteomalacia. Animal models of osteopetrosis provide evidence for osteoclast abnormalities in macrophage colony-stimulating factor (M-CSF), c-kit, c-fms, c-src, and c-fos, but the only genetic abnormality identified in human disease is a deficiency in carbonic anhydrase type II.

Marrow transplantation to replace the abnormal osteoclasts is the current mainstay of treatment. Clinical improvement is common and cures may be achieved for some patients. Drugs that can increase or support osteoclast function, such as calcitriol (high dose), M-CSF, and interferon-gamma1b (increase superoxide production), have had variable effects. Gene therapy holds promise for treatment upon identification of causative mutations.

Osteogenesis Imperfecta

Osteogenesis imperfecta (OI), also known as brittle bone disease, is an inherited disorder of type 1 collagen biosynthesis. The involvement of this most abundant protein in bone accounts for the major skeletal manifestations of the disease, but skin, ligaments, teeth, and sclera are also involved, with the latter two respectively accounting for the occurrence of dental disease (dentinogenesis imperfecta) and "blue sclera." The skeletal manifestations are recurrent fractures and skeletal deformity.

There are many classifications of this group of disorders, but they are all descriptive and overlapping. Classifications are also based on the mode of inheritance, which can be autosomal dominant or recessive. However, these classifications tend to obscure the clinical heterogeneity of the disease, which ranges from lethal to indolent.

Classification based on specific genetic defects is now beginning to emerge. Many specific defects of the genes that encode the pro-alpha 1 and pro-alpha 2 chains of type I collagen have been identified. These mutations interfere with the orderly array and mineralization of collagen and result in a structurally inadequate skeleton.

There is no effective treatment of OI, although calcitonin (CT) and bisphosphonates have been reported to be beneficial in some patients. Involved patients and their families should contact the Osteogenesis Imperfecta Foundation for up-to-date information.

Fibrous Dysplasia and McCune-Albright Syndrome

Fibrous dysplasia (FD) is a sporadic developmental bone disorder characterized by a fibrous expansion of bone-forming mesenchyme and bone cysts. Reminiscent of Paget's disease, FD can be monostotic or polyostotic. The skull, facial, and long bones are commonly involved. Deformity and fracture commonly result, and there is an increased incidence of sarcomatous transformation, perhaps contributed to by overexposure to diagnostic x-rays.

McCune-Albright syndrome (MAS) refers to the unusual combination of polyostotic FD with distinctive cafe-au-lait spots and endocrine hyperfunction. Pseudoprecocious puberty in girls is the most common endocrine disorder, but hyperthyroidism, Cushing's syndrome, acromegaly, hyperparathyroidism, hypophosphatemic osteomalacia, and hyperprolactinemia can also occur.

There is no effective treatment for FD, but CT and bisphosphonates show some promise. The skeletal disease does not seem to progress after age 40. Surgery can help the skeletal deformities, especially facial. An activating mutation of the stimulatory guanine nucleotide-binding protein (G_s) alpha subunit has been recently demonstrated in the involved skin and endocrine tissue in MAS.

Summary

Although the skeletal diseases described in this chapter are rare, they are important because of the basic and clinical insights they provide about the skeleton in general and the relevant mechanisms in specific. Osteopetrosis has provided revealing information about the cellular and molecular biology of the osteoclast. Osteogenesis imperfecta (OI) has demonstrated the importance of molecular biomechanical mechanisms for collagen in the structural integrity of the skeleton. Fibrous dysplasia (FD) and McCune-Albright syndrome (MAS) have revealed the role of G proteins in the growth and function of a variety of endocrine and skeletal tissues. Many of these diseases have provided gene targets for "knock out" and "knock in" animal models of skeletal disease.

These diseases are dramatic in their skeletal presentation but subtle in biochemical presentation, unlike the more common skeletal diseases discussed elsewhere. In general, these skeletal diseases have normal serum chemistries. The exceptions are the patients with osteopetrosis who have low blood calcium because of defective bone resorption and high serum acid phosphatase because of compensating osteoclasts and patients with MAS who have osteomalacia and hypophosphatemia. Awareness of the major clinical manifestations of these rare diseases can lead the astute clinician to the correct diagnoses and appropriate clinical management.

SUGGESTED READING

Anderson HC, Hsu HH, Morris DC, Fedde KN, Whyte MP. Matrix vesicles in osteomalacic hypophosphatasia bone contain apatite-like mineral crystals. *Am J Pathol*. 1997;151:1555-1561.

Astrom E, Soderhall S. Beneficial effect of bisphosphonate during five years of treatment of severe osteogenesis imperfecta. *Acta Paediatr*. 1998;87:64-68.

Bembi B, Parma A, Bottega M, et al. Intravenous pamidronate treatment in osteogenesis imperfecta. *J Pediatr*. 1997;131:622-625.

Bianco P, Kuznetsov SA, Riminucci M, Fisher LW, Spiegel AM, Robey PG. Reproduction of human fibrous dysplasia of bone in immunocompromized mice by transplanted mosaics of normal and Gsα-mutated skeletal progenitor cells. *J Clin Invest*. 1998;101: 1737-1744.

Charles JM, Key LL. Developmental spectrum of children with congenital osteopetrosis. *J Pediatr*. 1998;132:371-374.

Chen CC, Czerwiec FS, Feuillan PP. Visualization of fibrous dysplasia during somatostatin receptor scintigraphy. *J Nucl Med*. 1998;39:238-240.

Fedde KN, Michell MP, Henthorn PS, Whyte MP. Aberrant properties of alkaline phosphatase in patient fibroblasts correlate with clinical expressivity in severe forms of hypophosphatasia. *J Clin Endocrinol Metab*. 1996;81:2587-2594.

Glorieux FH, Bishop NJ, Plotkin H, Chabot G, Lanoue G, Travers R. Cyclic administration of pamidronate in children with severe osteogenesis imperfecta. *N Engl J Med.* 1998;339:947-952.

Jensen BL, Lund AM. Osteogenesis imperfecta: clinical, cephalometric, and biochemical investigations of OI types I, II, and IV. *J Craniofac Genet Dev Biol.* 1997;17:121-132.

Key LL Jr, Rodriguiz RM, Willi SM, et al. Long-term treatment of osteopetrosis with recombinant human interferon gamma. *N Engl J Med.* 1995;332:1594-1599.

Korkko J, Ala-Kokko L, De Paepe A, Nuytinck L, Earley J, Prockop DJ. Analysis of the COL1A1 and COL1A2 genes by PCR amplification and scanning by conformation-sensitive gel electrophoresis identifies only COL1A1 mutations in 15 patients with osteogenesis imperfecta type I. *Am J Hum Genet.* 1998;62:98-110.

Landsmeer-Beker EA, Massa GG, Maaswinkel-Mooy PD, van de Kamp JJ, Papapoulos SE. Treatment of osteogenesis imperfecta with the bisphosphonate olpadronate (dimethylaminohydroxypropylidene bisphosphonate). *Eur J Pediatr.* 1997;156:792-794.

Moriwake T, Seino Y. Recent progress in diagnosis and treatment of osteogenesis imperfecta. *Acta Paediatr Jpn.* 1997;39:521-527.

Redford-Badwal DA, Stover ML, Valli M, McKinstry MB, Rowe DW. Nuclear retention of COL1a1 messenger RNA identifies null alleles causing mild osteogenesis imperfecta. *J Clin Invest.* 1996;97:1035-1040.

Sarafova AP, Choi H, Forlino A, et al. Three novel type I collagen mutations in osteogenesis imperfecta type IV probands are associated with discrepancies between electrophoretic migration of osteoblast and fibroblast collagen. *Hum Mutat.* 1998;11:395-403.

Scott JE, Tenni R. Osteogenesis imperfecta mutations may probe vital functional domains (eg, proteoglycan binding sites) of type 1 collagen fibrils. *Cell Biochem Funct.* 1997;15:283-286.

Whyte MP, Walkenhorst DA, Fedde KN, Henthorn PS, Hill CS. Hypophosphatasia: levels of bone alkaline phosphatase immunoreactivity in serum reflect disease severity. *J Clin Endocrinol Metab.* 1996;81:2142-2148.

INDEX

Note: Page numbers in *italics* indicate figures; page numbers followed by t indicate tables.

14

14

14

14

14

14

14

14

14

14

14